Naturalists' Handbooks 23

Blowflies

ZAKARIA ERZINÇLIOĞLU
Department of Zoology
University of Cambridge

With illustrations by Sarah Bunker

Published for the Company of Biologists Ltd by

The Richmond Publishing Co. Ltd

P.O. Box 963, Slough SL2 3RS

Series editors
S. A. Corbet and R. H. L. Disney

Published by The Richmond Publishing Co. Ltd,
P.O. Box 963, Slough, SL2 3RS
Telephone: 01753 643104

ISBN 0 85546 303 1 Paper
ISBN 0 85546 304 X Hardcovers

Printed in Great Britain

For Sharon, my wife

Contents

Editors' preface

Blowflies have much to recommend them as subjects for investigation. For people with an active interest in biological studies, special interest attaches to groups of organisms that are common and readily available, large enough to be recognised by eye in the field, readily reared in the laboratory, and significant for human welfare. In all these respects, blowflies achieve a high score. Their larvae are familiar as bait for anglers and as the organisms responsible for sheep strike, and they sometimes play a part in solving murder mysteries. The adults are strikingly handsome. Though unwelcome in the kitchen they are rewarding subjects for ecological studies in the field, and their applied importance gives an added incentive. This book introduces the natural history of blowflies and draws attention to some unanswered questions about their biology. It offers illustrations and keys for identification, and describes some research methods that will help readers undertaking practical investigations to answer some of those questions. We hope this book will stimulate wider appreciation of the great biological interest of these distinctive flies.

S.A.C.
R.H.L.D.
March 1995

Acknowledgements

I would like to express my thanks to Dr Sally Corbet, Dr Lewis Davies and Dr Henry Disney for their interest and encouragement during the writing of this book. I am grateful to Professor Gabriel Horn and Dr Ken Joysey for providing facilities at the Department of Zoology and the Museum of Zoology, University of Cambridge. I benefited from discussions with Professor Bernard Greenberg, and with Mr Ken Smith who also read and commented on the draft keys. My particular thanks go to Sarah Bunker, whose illustrations enhance this book.

1 Introduction

1.1 Why bluebottles and other blowflies are worth studying

Blowflies are among the commonest and most familiar of all insects. They are found in tropical rain forests, deserts, oceanic islands, temperate lands, and even Arctic wastes, although they are absent from Antarctica. They are abundant throughout the British Isles. As a result, everyone is familiar with blowflies, which are known to most people as bluebottles and greenbottles. These insects, together with their lesser-known relatives, form the subject of this book.

Because of their abundance, and because they are easy to keep and breed in captivity, blowflies are ideal insects to study. They have attracted large numbers of research workers for several reasons. First, the larvae of some species invade the tissues of living mammals such as sheep, or even man himself. Investigation of 'sheep-strike' in particular has produced much information on blowflies. Secondly, because blowflies are easily cultured in the laboratory and their development is rapid, they are useful animals in which to investigate problems of insect physiology and cell function. Thirdly, blowflies very quickly discover a corpse and so their larvae may be useful to forensic scientists concerned with such questions as the time of death of the victim.

Many aspects of the biology of British blowflies remain to be explored, and the results of this research may be of value to those involved in the control of sheep-strike or forensic investigations. This book aims to encourage investigation of the many neglected aspects of blowfly biology. Before proceeding with an introduction to blowfly biology, a few words on the nature of scientific research may be useful.

1.2 Research

As stated above, one of the main aims of this book is to encourage the reader to conduct research on blowflies. It is unfortunate that a large number of people who are quite capable of conducting useful scientific investigations are deterred from so doing because of a number of erroneous beliefs.

The first is that important scientific discoveries require sophisticated equipment. This is not necessarily so. Much of the ecological research with which this book is concerned requires only a pen and notebook, and a few simple items of equipment such as are described in chapter 6. An example of an excellent piece of ecological work that utilised the minimum of equipment is that by Denno & Cothran (1975)* who used only a few tubes and other

* References cited under authors' names in the text appear in full in Further reading on p. 60

readily accessible items such as forceps and alcohol for preservation.

A second widespread belief is that any research that is worth doing must already be well in hand by professional scientists in universities and other research institutes, and that, therefore, the amateur cannot make any really useful or original contributions. This is simply not true, and many worthwhile discoveries are continually being made by serious amateurs, not only in biology, but in other fields of science as well.

Many amateurs believe that professional scientists work according to a mysterious system known as the 'scientific method', which endows their work with a rigour denied to the work of amateurs. Scientists do set about their work in a methodical and critical manner, but so do other people who are engaged in exploratory activities, the aim of which is to arrive at the truth. Scientists employ a mixture of observation and experiment in their work, and this is essentially the method employed by workers in a very wide range of other human activities, such as detectives, electricians or car mechanics.

The way one approaches a particular scientific problem depends largely on its nature; common sense must prevail. The question should be absolutely clear and formulated in such a way that an attempt can be made, by observation and experiment, to answer it. Often, in the early stages of research, we start with a question like, for example, "What do blowflies feed on?" After some initial observations, the question may be narrowed down to, for example, "Which of the known food substances is the preferred one?" A series of experiments may then be undertaken to answer this question. In this way, a continually deepening knowledge of a particular aspect of the biology may be gained.

The bibliography at the end of this book should lead the reader into the relevant literature. While the would-be researcher must read enough to be able to contribute usefully to the subject, too much reading cannot be a good thing. In the words of Sir Peter Medawar (1979):

*"Too much book learning may crab and confine the imagination.....The beginner **must** read, but intently and choosily and **not too much**".*

The book from which this quotation was taken is an excellent introduction to the nature of scientific research, and is strongly recommended.

1.3 What is a blowfly?

The class Insecta, to which all insects belong, comprises approximately twenty orders, of which the order Diptera is one. The Diptera includes all true flies. A true fly is an insect which has only one pair of functional wings (the forewings), the hindwings being modified to form club-like halteres. Although some other insects possess only one pair of functional wings, no others have the hind pair modified into the highly characteristic dipteran halteres. The order Diptera is one of the largest orders, containing over 100,000 described species. It is divided into three main sub-orders, the Nematocera, the Brachycera and the Cyclorrhapha (Table 1).

Table 1. *The three sub-orders of the Diptera, and some representative families*

NEMATOCERA	BRACHYCERA	CYCLORRHAPHA
Culicidae (mosquitoes)	Tabanidae (horseflies)	Calliphoridae (blowflies)
Chironomidae (midges)	Asilidae (robberflies)	Sarcophagidae (fleshflies)
Ceratopogonidae (biting midges)	Bombyliidae (beeflies)	Muscidae (houseflies)
Simuliidae (blackflies)	Stratiomyidae (soldierflies)	Tachinidae (parasite flies)

The blowflies comprise the family Calliphoridae in the sub-order Cyclorrhapha, which includes all the higher flies. The other families listed under the Cyclorrhapha have many similarities to blowflies, and will be referred to at various points in the book.

Colyer & Hammond (1968) give a well-illustrated introduction to the natural history and classification of British flies.

1.4 Life cycle

The female of a typical blowfly lays her eggs on the carcase of an animal, and the larvae hatch and feed voraciously on the decomposing flesh for a number of days and eventually pupate in the soil. After a variable period of time, the adults emerge and the cycle starts all over again. This highly successful life strategy, and its development over many millions of years of evolutionary history, will be explored in this book. A great many blowfly species have very different and often bizarre life-histories; in fact, the majority of species do not breed in carcases, but our knowledge of them is very scanty. There are many areas of the biology of blowflies where our understanding is only rudimentary. Perhaps this book will encourage the reader to investigate some of these points, and to contribute to the knowledge of these fascinating insects.

2 The biology of blowflies

2.1 Introduction

A great deal of laboratory-based research has been carried out on carrion blowflies, largely because they are easily kept and reared in the laboratory and are, therefore, useful experimental animals. Much less is known about their behaviour in nature, and it is this aspect of their biology that is our main concern in this book. This chapter describes what is known about the basic biology of blowflies, emphasising unexplored areas where the reader is encouraged to carry out original investigations. We deal first with those species that breed in carcases, and we then examine species with different life styles.

2.2 Searching for a carcase and laying eggs

A female blowfly carrying eggs (a gravid female) can detect the presence of a carcase over a remarkably great distance. This is a fact that everyone who has studied blowflies knows, and yet the actual distances involved do not seem to have been measured, at least for British species. MacLeod & Donnelly (1963) claimed that the greenbottle, *Lucilia sericata,* could be recruited to carcases up to 6.4 km away. But their flies may have been transported by vehicles over at least part of the distance measured. More recently, Braack (1986) showed that some South African blowflies of the genus *Chrysomya* were caught in baited traps up to 63.5 km away from a point of release of marked flies. The genus *Chrysomya* does not occur in Britain, and it would be interesting to see from how far away other blowflies, especially the common *Calliphora* (bluebottle) and *Lucilia* species, are recruited to a carcase. This might be done by releasing numbers of marked flies at known distances from a baited trap of a kind that will retain the visiting flies (technique, p. 53).

Blowflies detect carcases primarily by odour. Their sense of smell is very well developed; odours are detected by the hair-like sense organs on the third segment of the antenna. The physiology of these sense organs has been well studied in the laboratory; the elegant story of these experiments has been told by Dethier (1976).

The attractiveness of a carcase to blowflies varies with its degree of decomposition. My preliminary experiments showed that mouse carcases that were allowed to decompose for 10 days at a constant temperature of 18° C were much more attractive to *Calliphora vicina* than the carcases of freshly killed mice. In fact, caged *C. vicina,* when given a choice between the two kinds of carcase, completely ignored the fresh ones and laid eggs only on the decomposed carcases. Other species have been found to favour an earlier or later stage of decomposition in nature (Lane, 1975). Most blowflies can probably detect carcasses within a few hours after death, but carcasses that have passed a certain stage of decomposition, or are mummified or dry, are no longer attractive (Nuorteva, 1977). There is great scope for research, either in the laboratory or the field, to study the responses of blowflies to meat or carcasses at different stages of decomposition.

The size and position of the carcase are also known to affect its attractiveness to blowflies. Thus, *Calliphora vomitoria* seems to be preferentially attracted to larger carcasses (Hennig, 1950; Nuorteva, 1959; Davies, 1990), while *Lucilia richardsi* chiefly attacks small rodent or shrew carcasses (Nuorteva & Skaren, 1960). Carcasses lying in sunlight are generally more attractive to *Lucilia* than to *Calliphora* species, although certain woodland *Lucilia* species (such as *L. richardsi*) probably avoid direct sunlight. It has sometimes been suggested that *Lucilia* species will not lay eggs on carcasses whose temperature is below 30° C. This is not strictly true; these species will lay their eggs on carcasses at a much lower temperature in the laboratory. The situation in nature requires examination.

When observing the behaviour of blowflies on and around a carcase, I have often noticed that the flies first land somewhere nearby, before approaching and landing on the carcase itself. Carcasses are of interest to many kinds of animals, and it is possible that the flies' behaviour is an adaptation to avoid predation.

Although a wide range of natural substances are attractive to blowflies, only a few induce egg-laying. Generally speaking, the most attractive objects are the carcasses of vertebrate animals and other materials that contain the proteins albumin and globulin. There is evidence that the products of amino acid decomposition (decarboxylation) are very attractive to blowflies. Laboratory studies have shown that the antennae of blowflies respond to compounds with a chain-length of 5 to 7 carbon atoms and an alcohol, aldehyde or keto group (Kaib, 1974).

Decomposition odours that smell strongly to the human nose are not necessarily very attractive to blowflies. Thus, decomposing fish, considered a most offensive smell by most human beings, is only moderately attractive to many blowflies. In a series of field and laboratory experiments I found that *Calliphora vicina* preferentially laid eggs on small mammal carcasses and avoided fish carcasses when given a choice. Interestingly, the Tropical Asian species, *Chrysomya megacephala*, responded in exactly the opposite way in laboratory tests, laying preferentially on fish. It is intriguing to speculate on the origin of such preferences. There is evidence that the responses of insects to chemicals are influenced by their larval diet; an insect that fed on a particular substrate as a larva may be preferentially attracted to that substrate as an adult. This effect of larval experience on adult behaviour was once thought to result from a larval learning process called pre-imaginal conditioning. Recently, however, it has been proposed that traces of chemical cues are retained by the larva, and are bequeathed to the adult as a 'chemical legacy' (Corbet, 1985). These cues, rather than remembered larval experience, would mould the adult's behaviour. Whatever the exact mechanism may be, it would be illuminating to study the effects of larval diet on the subsequent chemical choice behaviour of adult blowflies.

Other factors may affect the attractiveness of a carcase. For example, it was found that females of the Australian species *Lucilia cuprina* were attracted to lay on carcases on which other females were already present and laying eggs (Barton Browne and others, 1969). This behaviour was found to be at least partly mediated by pheromones (chemical messenger substances which are secreted externally by insects and to which other members of the same species respond). Resulting as it does in large masses of eggs laid close together, this behaviour may minimise water loss in the egg-mass as a whole. The presence of eggs or larvae on a carcase may also affect its attractiveness to adult blowflies, although my own experiments seem to indicate that the odour produced by larval activity neither attracts nor repels British blowflies under natural conditions. In these experiments, equal weights of fresh liver, and liver on which a known number of larvae had fed for 24 hours before being removed, were exposed to blowflies (*Calliphora vicina*) in the field. Eggs were laid on both kinds of liver, and there was no significant difference in egg numbers between the two kinds of liver. On the other hand the Australian species *Lucilia cuprina* has been found to be attracted to volatile compounds extracted from media infested with larvae and micro-organisms; the species of larvae did not make any difference to the attractiveness of the volatiles, odours produced in the presence of other calliphorid and sarcophagid species being equally attractive. Flies were not attracted to volatiles extracted from media infested with micro-organism-free larvae (Eisemann & Rice, 1987). There is scope for further work here, especially on the possible effect of the actual presence of larvae (of different stages) and eggs.

It is sometimes said that blowflies do not lay eggs in the dark. This is not strictly true. They do not seem to lay eggs at night, but some will enter very dark places, such as cellars and chimneys, in search of carcases during the daytime when they are normally active.

It is easy to recognise a blowfly that is ready to lay eggs (oviposit). It flies with a characteristic buzzing noise, and has its abdomen swollen with eggs, and very often its ovipositor (the telescopic tube through which the eggs are laid) extruded. Having located a carcase, a female blowfly does not normally lay eggs immediately, but roams over the carcase for a while, apparently seeking a suitable place. The usual places selected for oviposition are the natural body openings (mouth, anus, eyes, ears) and any wounds that may be present. Frequently, eggs are laid underneath the body. The nineteenth century entomologist Fabre suggested that the preferred site of oviposition was the head, and recommended the placing of paper hoods over the heads of blackbird carcasses, a French delicacy, in market stalls as a measure against blowfly attack! It is my belief that whole carcases are much more attractive to blowflies than butchered meat. Whole carcases are less likely to dry up, and would, therefore, be a more suitable environment for the eggs and resulting larvae. This hypothesis needs testing.

We have seen that dry carcases are generally unattractive to blowflies, and it is also known that, in the Australian species *Lucilia cuprina*, females will not lay eggs unless their feet have made contact with free water (Barton Browne, 1962). This is clearly an adaptive feature, since on a dry carcase the eggs might dry up, or the hatching larvae might fail to penetrate and feed on the tissues.

Blowflies can lay very large numbers of eggs. A female *Calliphora vicina* can produce 2000–3000 eggs during its lifetime (Hinton, 1981). Up to 180 eggs may be laid during one bout. This is a necessary strategy for the blowfly since, as we shall see later, the eggs and larvae suffer from the depredations of a wide range of parasites and predators, and few of the eggs will survive to produce adult flies.

Fly oviposition activity is largely determined by temperature, and it is often stated that such activity ceases below 12° C. In fact, certain species will lay at lower temperatures (Erzinçlioğlu, 1986*b*), and further work is needed in this field.

The extent to which blowflies can detect buried carcases is unknown. It has been suggested that blowflies cannot detect a carcase covered by a layer of soil more than 2.5 cm deep, but this is likely to depend on the conditions. Does the soil type influence the flies' behaviour? Is there any difference in the response to loosely and tightly packed soil? Does the moisture content of the soil matter? Do the flies detect the carcase but fail to lay eggs? Do all species behave in the same way? These questions are important to forensic scientists, and deserve detailed investigation.

People often ask whether carrion blowflies will lay eggs on substrates other than carcases. Some exotic carrion blowflies will certainly oviposit on dung. *Chrysomya albiceps* is an example (Erzinçlioğlu & Whitcombe, 1983), but there is no definite record of this for any British species, although Bogdanov (in Haddow & Thomson, 1937) claimed to have reared *Calliphora vicina* on dung in the laboratory. In the former Soviet Union *Lucilia illustris* breeds in mink dung and *Phormia terraenovae* breeds in soil contaminated with poultry urine and droppings. From time to time one hears of carrion blowflies being reared from exclusively vegetable material (although a number are known to be associated with fungi), but there appear to be no definite records in the literature for this. The New Zealand species *Ptilonesia auronotata* has been thought to breed in seaweed on the sea shore, although this is now in doubt (Dear, 1986).

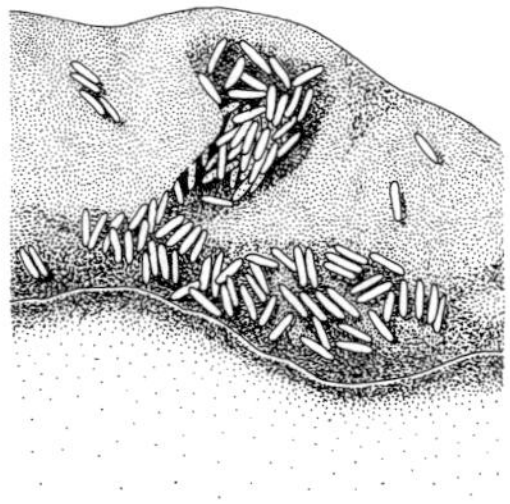

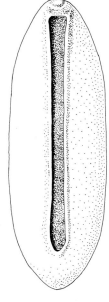

Fig. 1. Eggs of the blowfly
Calliphora vomitoria on meat.

2.3 The eggs

The blowfly has now laid its eggs and departed; it will take no further interest in them. A blowfly egg is white or pale yellow, and cylindrical, tapering slightly at each end (fig. 1). It measures about 1.5 mm by 0.4 mm, although the size will vary between species. The egg is slightly curved like a banana, with the upper side (in relation to the body axis of the egg-laying fly) concave, the lower side convex. The egg-shell, or chorion, is covered over most of its surface with a hexagonal pattern. Along the mid-line of the upper surface, however, the chorion has a rather different structure. Here, it forms a groove with a raised ridge on either side (fig. 2). The 'median area' between the ridges has a spongy, porous structure, and is the respiratory area of the egg. When the egg is submerged, as often happens when it rains, this area is not wetted. It retains a film of air, and can act as a plastron, or physical gill. The ridges are the hatching pleats along which the chorion splits when the larva emerges, the median area peeling off from the front to the hind end. At the anterior end of the egg is a small hole, the micropyle, through which the sperm passes to fertilise the egg.

The eggs usually take a day or two to hatch, depending on temperature, humidity and species. Hatching occurs soonest at around 35° C for *Lucilia* species, but the temperatures above which eggs will not hatch do not seem to have been determined. At low temperatures the eggs take much longer to develop. It has been said that those of *Calliphora vicina* will not hatch at, or below, 4° C (Nielsen & Nielsen, 1946). This is a point worthy of further investigation in this and other species. Oxygen deficiency in the water film around the egg will also slow down development (Anderson, 1960), and low humidities make hatching difficult, because of changes in the shape of the chorion (Davies, 1950).

Hinton (1981) gives a richly illustrated and comprehensive account of insect eggs, including those of blowflies. Vogt & Woodburn (1980) give much information on the effects of temperature on egg development.

Fig. 2. An egg of the blowfly
Calliphora.

instar: the developmental
stage of an insect between
successive moults

2.4 The larval stages

When it hatches, the first-instar larva is small and whitish. This larva is easily dehydrated if conditions are a little dry, and easily drowned if too moist; it seems to be the most vulnerable stage in the entire life cycle. On hatching it immediately moves to a part of the carcase where conditions are suitable. The first instar is a short-lived stage; on average at room temperature, it will moult to the second instar in a little over a day. The second instar will last for a similar period, before moulting to the third instar. This is the well-known fisherman's maggot (fig. 3). It feeds voraciously for 3 or 4 days, increasing its weight and size by up to 6 or 7 times.

Like the other instars, the third instar is a 12-segmented, white or yellowish-white maggot. On the second

Fig. 3. A blowfly larva (head to the left).

(a)

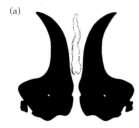

(b)

oral sclerite

mouth
hook

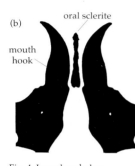

Fig. 4. Larval cephalo-
pharyngeal skeleton, seen from
below, showing the oral sclerite
between the two mouth hooks
(a) *Lucilia sericata*
(b) *Calliphora vicina.*

and twelfth segments are two pairs of spiracles, the respiratory openings that lead to the tracheal system (the system of interconnecting tubes that supply oxygen to, and remove carbon dioxide from, the tissues). At the front end of the larva, internally, lies a complicated structure, the cephalopharyngeal skeleton, which is the anterior part of the alimentary canal. It varies in structure between instars and between species (fig. 4a, b). The morphology of blowfly larvae is dealt with by Erzinçlioğlu (1985), and Roberts (1972) describes the way the mouthparts work.

Blowfly larvae, like all known cyclorrhaphan (higher fly) maggots, digest their food externally by releasing enzymes into the surrounding food. It has been suggested that this habit was a major step in the evolution of flies (Disney, 1986); it is coupled with the presence of mouth-hooks (fig. 4) that act as grapnels to pull food into the mouth. These are quite unlike the opposable mandibles of the more primitive fly larvae; such mandibles are used to bite off pieces of food.

Blowfly larvae feed constantly, and are usually deeply embedded in the decomposing flesh. Light is probably the main factor that stimulates the larvae to burrow; by moving away from light they move deeper into the carcase (Patten, 1914). A positive attraction to the carcase must also play a part, but the olfactory responses of maggots to the smell of meat appear not to have been investigated, and would be a fruitful line of research. Blowfly larvae produce ammonia in large amounts as an excretory product. Their responses to this substance, amongst others, could usefully be investigated. The article by Bolwig (1946) on the olfactory behaviour of housefly larvae should be useful when designing experiments in this field.

The quality and quantity of the larval diet has a profound effect on the size and longevity of the adult fly, and on the number of eggs it will produce (its fecundity). In laboratory cultures, inadequately-fed larvae develop into stunted flies. More surprisingly, larvae of *Calliphora vicina* that have been fed on fatigued (exercised) frog muscle can attain a body weight 9% greater than those fed on rested muscle (Munro Fox & Pugh Smith, 1933). Certain bacteria and yeasts seem to promote larval growth, since *Lucilia sericata* larvae developed more slowly when reared on sterile muscle (Hobson, 1932). Interestingly, however, blowfly maggots secrete a substance toxic to certain strains of bacteria (Gwatkin & Fallis, 1938), and this is probably why maggot-infested carcases do not support such large colonies of bacteria and fungi as uninfested ones. Another unusual effect of larval diet concerns the sex ratio. Some *Lucilia sericata* larvae were allowed to feed until they left the meat, while others were allowed to feed for only 30–36 hours as larvae. The flies that emerged from the first group showed a sex ratio of 28%–31% males to 69%–72% females, while the flies from the second group showed a ratio of 62%–65% males to 35%–38% females. The ecological significance, if any, of this phenomenon is unknown (Herms, 1928).

Little is known of the effect of species of carcase on larval development, although, as we saw earlier (section 2.2), certain fly species seem to lay eggs preferentially on particular kinds of carcase. A comparison of the effect of two different carcase species on rate of development, adult size, sex ratio, longevity and, particularly, fecundity would be of great value.

Temperature is a major factor affecting the rate of larval development. Developmental rates at various constant temperatures have been determined for a number of species (see Smith, 1986; Nuorteva, 1977; Williams, 1984; Williams & Richardson, 1984; and especially Reiter, 1984), but little information is available on the effects of fluctuating temperatures. The evidence so far suggests that fluctuating temperatures retard development, increasing the longevity of the maggot (see Greenberg, 1991).

It is my strong impression that larvae that are reared at lower temperatures produce larger adults. This idea has been explored in mayflies (see Corbet and others, 1974) and some other insects, but has not been investigated in blowflies, and is worth looking into.

The larval mass is usually a few degrees warmer than its surroundings, presumably because of the metabolic heat produced by the larvae. Dr Bernard Greenberg has suggested that the larvae may regulate their temperature by moving closer to one another at low temperatures, and away from one another at high temperatures. In this way larvae might keep warm enough to grow fast, while avoiding overheating.

The remarkable phenomenon of paedogenesis, or reproduction by larvae, is known to occur in certain flies. Its occurrence in blowfly larvae has never been adequately demonstrated, and has been refuted by Keilin (1924). It has since been discovered in other higher flies, the hoverflies, by Ibrahim & Gad (1978), and may yet be found in blowflies.

2.5 Pupation

When the larva has finished feeding, it empties its gut and leaves the carcase, usually at night. It is easy to distinguish a larva that has not finished feeding from one that has, since the gut contents can readily be seen through the cuticle. Changes occur in the morphology of the crop when feeding is completed, and these can be used to age the larva (Greenberg, 1985). A larva that has finished feeding (a pre-pupa) wanders away from the carcase before burrowing into the soil. Larvae may wander up to 6.4 m over soil (Cragg, 1955), and over 30 m on the concrete floors of slaughterhouses (Green, 1951). Not all species migrate, however; *Phormia terraenovae* usually pupates on the carcase.

In most species, pupation occurs in the top 2 or 3 cm of soil. The cuticle of the larva contracts, hardens and darkens to form the puparium (pls. 3.9 and 3.10), which is the outer protective covering of the pupa. In blowflies and other higher flies, the pupa therefore has two coverings, its own cuticle and the puparium, or modified third-instar larval cuticle. It is important not to confuse the pupa, which is the living insect at a particular developmental stage, with the puparium, which is the non-living structure that encloses the pupa. Fraenkel & Bhaskarian (1973) give details of pupation and puparium formation.

When the puparium forms, the tracheal connections of the third-instar spiracles are broken, although they remain visible as external structures of the puparium. However, they are not used by the pupa, which has its own respiratory structures. When the cuticle hardens and darkens, two small, round patches on the upper surface of the fifth segment remain soft and pale; each of these is eventually pierced by a thumb-like structure, the respiratory horn of the pupa (pl. 3.9). (The horns appear to be borne by the fourth segment of the puparium; this is an illusion due to the fact that the first segment contracts into the second during puparium formation, so that the fifth segment appears to be the fourth.) Keilin (1944) describes these and many other respiratory adaptations in fly larvae and pupae.

The developmental rate of the pupae, like that of the larvae, is very much a function of temperature. Dallwitz (1987) compared the effects of constant and fluctuating temperatures on pupae of the Australian *Lucilia cuprina*. He found that the pupae developed faster at a constant temperature of 30° C than they did at a fluctuating temperature with a mean value of 30° C. Similar experiments on larvae might give a different answer, because the larva is a mobile stage and may be able to regulate its temperature (section 2.4).

The larva and the adult fly are very different organisms morphologically, ecologically and physiologically. The dramatic transformation from one stage to the other takes place in the pupa. Perhaps it is this that has enabled insects with a complete metamorphosis, such as flies, beetles and butterflies, to become so successful. Each stage can invade and exploit an entirely different habitat. Hinton (1948) discusses the origin and significance of pupation.

2.6 Adult emergence

When the pupa has completed its development and the adult is formed, the fly will emerge, shedding the puparium and also the pupal skin, which may be seen as a thin, greyish-white membrane adhering to the inner wall of the empty puparium. During emergence, the tip of the puparium (the operculum) splits off and breaks into two parts (or 'caps') (pls. 3.9 and 3.10). The uppermost cap bears the respiratory horns; the lower one bears the third-instar cephalopharyngeal skeleton internally.

By expanding and contracting a blood-filled sac, the ptilinum, on its head (fig. 5), the adult forces off the operculum, escapes from the puparium and digs its way upwards through the soil. The direction of digging is determined by light, not gravity (Fraenkel, 1935). Soon after

ptilinum

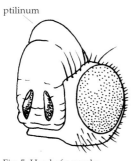

Fig. 5. Head of a newly-emerged blowfly, with the ptilinum expanded.

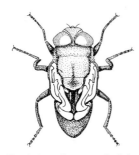

Fig. 6. A newly-emerged adult of *Calliphora vicina*. The wings have not yet expanded.

emergence, the ptilinum is withdrawn into the head and is never used again. A full account of the structure and function of the ptilinum is given by Laing (1935).

The newly emerged fly is a pale and crumpled creature (fig. 6), and it cannot fly because of its crumpled wings. It usually takes about 24 hours for the fly to become fully expanded and pigmented. If, for some reason, the duration of burrowing through the soil is prolonged, expansion and darkening are delayed (Fraenkel, 1935). While expanding and darkening, the flies usually conceal themselves in crevices, presumably to avoid predators. Shortly after emergence, the fly ejects a greenish-black fluid through its anus; this is the meconium, a mixture of products of pupal metabolism.

2.7 Adult feeding

The fully-expanded fly is now ready to feed. Blowflies need to feed for two reasons. First, in order to carry out their activities, they must ingest energy-rich foods. These they obtain mostly from nectar, although they are also known to feed on the honeydew of aphids (greenfly and relatives) (Grinfel'd, 1955), and rotting fruits and other sweet substances, such as jam, are also attractive. Since blowflies depend heavily on flowers for their energy intake, it would be useful to investigate their preferences in terms of colour, size, shape, position and species of flower. Do different species have different preferences? Flies in general seem to have colour preferences as judged by the numbers found in different coloured water traps. White and yellow seem to be particularly attractive (Disney and others, 1982). White flowers (and white paints) that look alike to us are of two different types to the eye of an insect which can see ultraviolet light; some reflect ultraviolet, and others do not. In studies on water traps of different colours, Kirk (1984) distinguished these two types of white, and found that most flies prefer whites that do not reflect ultraviolet.

The second major dietary requirement is particularly important for females. In order to mature her eggs, a female carrion blowfly must have a protein meal. This is often obtained by feeding on carcases, although other materials, such as dung, generally much more common than carrion, are also used as sources of protein. Males also feed on carrion, but, at least in some Australian species, protein does not seem to be necessary for maturation of the spermatozoa (Mackerras, 1933). Also, females of *Lucilia sericata* increase in weight faster when they are fed on protein; the same is not true of males (Evans, 1935).

Some blowfly species are obligate parasites, in other words, they lay their eggs on living hosts and when they hatch the larvae feed on the living tissues; they cannot develop in dead carcases. The females of such species can mature their first batch of eggs without having to take a protein meal. This phenomenon is known as autogeny and the fly is said to be autogenous. Most blowfly species do

autogeny: the ability of an adult female to mature her first batch of eggs without taking a protein meal need a protein meal to mature the first egg batch (they are anautogenous), but in some species, such as *Lucilia cuprina,* a number of autogenous individuals occur in wild populations. Experiments have shown that such individuals die sooner than anautogenous individuals when deprived of water, although the reverse was true when sucrose was withheld (van Gerwen and others, 1987).

The blowfly proboscis is a complex and delicate instrument for lapping and sucking (fig. 7). Liquefied food passes into a series of channels on the two broad lobes at the end, and from there into the mouth opening and up the proboscis. A full account of blowfly mouthparts is given by Graham-Smith (1930).

2.8 Reproductive behaviour

Species differ from one another in the period that elapses between emergence and mating. The females of some blowflies, such as the Australian *Lucilia cuprina,* do not seem to mate until their ovaries are mature (Barton Browne, 1958), and the female becomes more receptive to the male after a protein-rich meal (Barton Browne and others, 1980). Little is known about the timing of mating in British blowflies, although Parker (1968) has shown that mating takes place on the second day after emergence in *Phormia terraenovae.* Campan & Langa (1981) showed that female receptivity affects male behaviour in *Calliphora vomitoria.* In the species that have been studied, it seems that females mate only once, whereas males mate repeatedly. Males are often seen to mate with females that are feeding at carcases or other protein sources. An American species, *Cochliomyia hominivorax,* is known to respond sexually to certain components of the commercial attractive bait, 'Swormlure' (Broce, 1980). Blowfly courtship behaviour is simple; the male runs rapidly towards the female, usually approaching from the side, then mounts her, orients and mates.

It is usually relatively easy for a blowfly to find a protein meal, but it may be harder to find a place to lay eggs. If this takes a long time, the viability of the unlaid eggs may decline (Norris, 1965). In certain species of *Lucilia,* this happens after a period of 5 to 10 days. Little is known of the species of *Calliphora* in this regard.

When blowflies mate, sperm passes through the vagina and up the spermathecal ducts into the spermathecae. When the eggs are being laid, they pass singly down the oviduct and are fertilised by sperm coming down the spermathecal ducts. If a carcase is available, the fly deposits its eggs immediately, and they will continue their development on the carcase. However, if a carcase is not immediately available, the first egg that passes down the oviduct is fertilised and then held in the vagina until the fly finds a carcase. Since it has been fertilised, the egg continues to develop, and when laid it will be at a more advanced stage of development than all the eggs laid later. This is why one sometimes sees a single first instar larva amongst a mass of

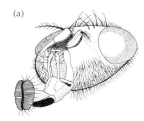
(a)

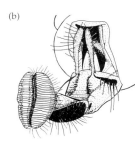
(b)

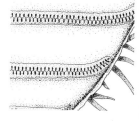
(c)

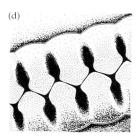
(d)

Fig. 7. The head and proboscis of *Calliphora vomitoria,* with successive enlargements (a, b, c, d) to show the fine channels into which liquid food is taken up.

unhatched eggs. I have known this to happen in *Calliphora vicina*, and it would be interesting to find out whether other species can hold more than one fertilised egg in the vagina. Female fleshflies, in the related family Sarcophagidae, retain their eggs in the body until they hatch, and give birth to larvae.

Many insects that breed in a specialised habitat restricted to a particular time of the year, such as the flower of a particular plant species, have only a single generation a year. Most carrion blowflies, however, go through several generations a year, since their breeding medium is available continuously. Of course, their breeding season is restricted by other factors, the most important of which is low temperature. In very cold climates, carrion blowflies may have only one generation a year. For example, this is true of *Calliphora alpina* in sub-arctic Finland. In Britain, most species probably have several generations a year, but little is known of the behaviour of such cold-adapted species as *Phormia terraenovae*, *Calliphora uralensis* and *Cynomya mortuorum* in the colder parts of the country, such as the mountains of northern Scotland.

2.9 Overwintering

Adult blowflies, like most adult insects, are warmth-loving animals, and it is natural to wonder where they go in the winter.

Some British blowflies, such as *Lucilia sericata*, overwinter as late third-instar larvae or pre-pupae. In some species of *Lucilia* and other genera, the pre-pupal period is prolonged in unfavourable conditions. This delay is not simply quiescence due to low temperature; it is a form of arrested development with a more complex physiological cause, known as diapause. It can be induced by unfavourable conditions acting either directly on the larva, or on the mother fly. In *L. sericata* the most important direct influence on the larva is a decrease in its moisture content; larvae that have been feeding on a dry carcase tend to undergo diapause when they leave it (Mellanby, 1938). Maternal induction of diapause has been shown to occur in *L. caesar*. In this species, the larval offspring of a female that has been subjected to shorter periods of daylight will enter diapause; it is the stimulus received by the mother that induces larval diapause (Ring, 1967). Maternal ageing also tends to induce larval diapause in the progeny.

Although diapause enables the species to overwinter at a suitably cold-tolerant stage in its life cycle, it is not the cold in itself that induces diapause. However, a period of cold is essential for the breaking of diapause. If a diapausing larva is subjected to low temperatures for several weeks and then brought back to a temperature high enough for development, it will end its diapause and resume normal development, whereas if a diapausing larva is kept at high temperature it will break diapause much later, if ever.

Blowfly larvae are remarkably tolerant of low temperatures. My experiments have shown that all larval instars of *Calliphora vicina* can survive repeated subjection to -6° C for 12 hours at a time. When brought back to room temperature these larvae continued to develop. This tolerance would seem to be an ecological necessity, since, during the winter, the temperature of the soil in which the larvae are overwintering may fall well below freezing. Block and others (1990) give further information on tolerance to freezing temperatures. However, Davies (1929) has shown that overwintering larvae of *Lucilia sericata* can avoid extremely low temperatures by moving vertically in the soil. Some species are better able to survive winter temperatures than others; Cragg (1956) has shown that larvae of *Lucilia sericata* suffered higher mortalities than those of *Calliphora vicina* when exposed in soil at an altitude of 555 m above sea level.

Not all blowflies are known to undergo diapause. Southern British populations of *Calliphora vicina* do not (Graham-Smith, 1916), whereas northern British populations do (Saunders, 1987). Work in the Soviet Union by Zinovjeva (1980) has also shown that northern populations undergo diapause and southern ones do not; this trait is under genetic control. Non-diapausing species do not necessarily overwinter in the pre-pupal stage but may do so in any stage that happens to be developing at the time the winter temperatures fall low enough to slow down development. *Calliphora vicina* adults may often be seen basking on walls on sunny winter days. Are these overwintering adults, or are they flies newly emerged from pupae? The situation requires examining. There does not seem to be any reference in the literature to the possibility of overwintering by the eggs, but this deserves attention because the eggs of *Calliphora vicina* can survive temperatures as low as -15° C.

Some species seem to overwinter mainly as adult flies. *Pollenia rudis* habitually overwinters in large numbers in attics of houses, and is often seen indoors in warm weather in the winter; this habit has earned it the name of cluster fly. *Protocalliphora azurea* is another species that overwinters as an adult. *Phormia terraenovae* is known to exhibit a true adult diapause, and overwintering adults were found to have accumulated large quantities of stored food reserves in the fat body.

2.10 Adult longevity

Overwintering adult blowflies may live for over six months, or even a year. Flies that are not overwintering do not live so long. Blowflies may live for up to six weeks, or even two months, in laboratory cultures. According to Norris (1965), the record for longevity outdoors is held by the American screw-worm fly, *Cochliomyia hominivorax*, an individual of which survived for 76 days in an outdoor cage. Little is known of the life-span of British species in the wild.

2.11 Natural enemies

Blowflies are subject to attack by a wide range of predators and parasites. The adults form part of the diet of many vertebrates such as birds, frogs, toads and lizards. According to Dr Miriam Rothschild, some birds prefer *Calliphora* species, finding *Lucilia* species less palatable. Robberflies and spiders doubtless also take their toll. The larvae are probably subjected to much heavier predation than the adults. This is because they cannot fly, they are present in locally high concentrations, and they live in a medium which is itself a favoured food of many animals. Their main predators seem to be beetles, particularly species of the genera *Carabus* and *Hister; Hister* beetles have even been proposed as blowfly biological control agents (Nuorteva, 1970). Other fly larvae are also known to prey upon blowfly maggots (section 2.18, p. 33).

Vertebrate predators on larvae include hedgehogs, which are sometimes found to have their stomachs full of maggots, and birds such as crows and choughs. It has been suggested that choughs are actually attracted to the maggots in carrion, rather than to the carrion itself; this may also be true of other birds. It is probable that many more birds prey upon maggots than is generally realised. Wobeser & Galmut (1984) studied the rate of digestion of blowfly maggots by mallards. They found that most maggots were unrecognisable, even as little as 15 minutes after being eaten. This rapid digestion makes it unlikely that maggots would be detected during a field necropsy except in birds that had fed during the previous 15 minutes. This may explain why maggots are hardly ever found in the guts of birds; most records of bird predation on maggots are based on observations of the birds' behaviour.

Adult blowflies do not seem to suffer from many parasitic diseases, although fungi of the genera *Empusa* and *Entomophthora* are common blowfly parasites. They invade through the soft membrane along the sides of the abdomen, and can grow very fast. Their attacks seem to be particularly common in the autumn, especially in wet weather. I have seen over 200 blowflies (mainly *Calliphora vicina,* but also other species of *Calliphora*) succumb to fungal attack after laying eggs on a gull carcase in a wood at Malham Tarn, Yorkshire, during September. Dead blowflies that have been killed by the fungus may sometimes be seen clasping the tip of a leaf or a blade of grass. Indoors, such flies may be seen stuck to window panes, the fungal spores distributed in a characteristic pattern around the fly, as a result of the fly's wing beat. Various thread worms (nematodes) are known to attack blowflies, but these seem to have been little studied.

A number of parasitic wasps attack the larvae of blowflies. Of these, the most important are the braconid *Alysia* species (fig. 8) and the chalcid *Nasonia vitripennis* (fig. 9). *Nasonia* is a very small insect, not easily observed in the field. Its larva is an external parasite of the pupa within the

Fig. 8. The braconid wasp *Alysia.*

Fig. 9. The chalcid wasp *Nasonia.*

Fig. 10. A puparium of
Calliphora with the exit hole of
the chalcid wasp *Nasonia*.

puparium, and it can doubtless cause high mortalities in
blowfly populations. In the laboratory, an attack by *Nasonia*
can mean the loss of entire cultures; instead of the expected
emergence of blowflies, one may find a cage full of these
minute parasites. On close examination, each puparium will
be seen to be perforated by a small, neat, round hole through
which the emerging wasps left their host (fig. 10). Laboratory
experiments have shown that when potential host pupae are
densely aggregated, a lower proportion of individuals are
parasitised (Jones & Turner, 1987), and it would be
interesting to see if the same happens in nature. A good
review of the biology of *Nasonia vitripennis* is given by
Whiting (1967).

 Alysia is a much larger insect, and can quite easily be
observed in the field, where it may often be seen fussing
about on larval masses in carcases. Unlike *Nasonia,* it is an
internal parasite of the pupa, but like *Nasonia,* it seems to be
catholic in its choice of host. Myers (1929) stated that,
although *Alysia manducator* is a parasite of *Calliphora vicina,* it
does not parasitise *Calliphora vomitoria*. This statement has
often been repeated in the literature, but it is not true; Evans
(1933) has reared *A. manducator* from *C. vomitoria* in the
laboratory, and I have recorded it attacking this species in the
wild. Parasitism by *Alysia manducator* is known to delay
puparium formation (Chernoguz, 1984). The biology of
Alysia manducator has been described in detail by Evans
(1933) and Salt (1932). Little is known of its effect on blowfly
populations in the wild, but Graham-Smith (1919) has
recorded parasitisation rates of 61% in *Calliphora vicina* in the
autumn. An excellent introduction to parasitic insects is
given by Askew (1978); Gauld & Bolton (1988) give much
information on parasitic wasps.

 Nematodes of the genera *Heterorhabditis* and
Steinermema are known to infect the larvae of the Australian
Lucilia cuprina, and it was found that rates of parasitism were
lower in soils of high clay content (Molyneux & Bedding,
1984). Hardly anything is known of the nematode parasites
of British blowfly larvae, and a rich field awaits the
investigator.

 Blowfly laboratory cultures sometimes succumb to a
curious infection, the causative organism of which does not
appear to be known. After feeding and leaving the carcase in
a seemingly healthy condition, the larvae contract, turn a
golden-brown colour, and die. I have known this to happen
with cultures of *Calliphora vicina, Calliphora vomitoria* and
Cynomya mortuorum although I know of no record of this
infection in the wild. The pathogen may be a bacterium or a
virus, but this remains to be shown.

2.12 Parasitism and predation by blowflies

 Blowflies can themselves act as parasites and
predators. The most widespread kind of parasitism by
blowflies is that which results in the disease known as
myiasis, where the tissues of living mammals (including

man) are invaded by maggots. This is a very unpleasant condition for the host, and is often fatal. The species that cause myiasis are of two kinds: facultative parasites, which normally develop in carrion, but sometimes parasitize living hosts, and obligate parasites, which cannot develop in carrion, but need a living host in order to complete their life cycle. There are no obligate parasites of mammals in the British blowfly fauna. Facultative parasites may be further subdivided into primary agents (those that can initiate a lesion, or invade a wound, in a previously uninfected animal) and secondary agents (those that invade only after parasitization has been initiated by another species). The term 'true myiasis' will here be used to refer to the invasion of living tissues by maggots, to distinguish it from other forms of blowfly parasitism, like blood-sucking.

Most agents of myiasis attack their hosts by laying eggs on wounds, but some, especially certain *Lucilia* species, lay eggs on the unwounded tissues of sheep and other animals. This important and widespread form of myiasis, known as 'sheep strike', resulted in great losses to sheep farmers in Britain until relatively recently, and many farmers dock the tails of their sheep to reduce the risk of strike (French and others, 1994). Treatment of sheep by dipping has helped to reduce the incidence of strike in Britain, although it is still a major problem in other countries, especially Australia and South Africa. Sporadic cases continue to occur in Britain, where myiasis in animals other than sheep is also known, although it has been little studied. Most of the extensive research on the ecology of myiasis has been carried out in relation to sheep strike.

The many factors that induce a blowfly to lay eggs on a living host seem to interact in an exceedingly complex manner. In spite of much research, the problem is far from being understood, and different workers have occasionally reached mutually contradictory conclusions. Major factors include odours and climatic conditions. As we saw above (section 2.2), the odours that attract blowflies to feed are not the same as those that elicit egg-laying. The attractants given off by potential hosts are probably often a mixture of both these sets of stimuli.

The odours given off by the clean, dry fleece of sheep attract female blowflies, especially if these are fertilised, but they do not stimulate them to lay eggs. (Male flies are not attracted.) Even fleece that has been washed in hot soapy water and ether retains its attractiveness to the female blowflies. The strength of attraction to wool differs between species, and species which are efficient myiasis agents are more strongly attracted (Cragg & Cole, 1956). Unlike clean fleece, fleece which has been soiled with urine does stimulate blowflies to lay eggs (Beveridge, 1935); perhaps this is due to odours resulting from the interaction of the urine with the fleece and the skin, or to the increased humidity. The non-British *Lucilia cuprina* will lay eggs on wet wool (Mackerras & Mackerras, 1944). Sheep excreta do not attract blowflies to lay eggs, but blowflies were found to lay eggs on sheep which had excreta experimentally placed upon them (Hobson, 1935). Lambs with faecal staining on their fleece are known to be particularly susceptible (Heath, in Dear, 1986).

Sweat might be a blowfly attractant, but there is little evidence for this. It has been suggested that the discoloured fleece of some Australian sheep might have a higher content of suint (dried sweat) than normal coloured fleece (Holdaway & Mulhearn, 1934), but in Welsh sheep there appears to be no correlation between discolouration and suint content, although there does appear to be a correlation between discolouration and sheep strike (Hobson, 1936). *Lucilia sericata* and the non-British *Lucilia cuprina* were found not to be attracted to preparations of suint in a series of experiments (Hepburn & Nolte, 1943).

Sheep which are already injured or diseased are frequently attacked by blowflies. Fleece-rot, caused by the bacterium *Pseudomonas aeruginosa*, makes sheep very susceptible to blowfly attack, and bacterial extracts from such sheep stimulate *Lucilia cuprina* to lay eggs (Emmens & Murray, 1982, 1983). The idea that volatiles released by bacterial decomposition attract blowflies to lay eggs is further supported by the finding that meat inoculated with a mixed culture of bacteria from sheep intestines is very attractive to *Lucilia sericata* and *Lucilia cuprina* (Hepburn & Nolte, 1943). Carrion, where active bacterial decomposition takes place, is much more attractive as an egg-laying site than living hosts to most blowfly species (other than obligate parasites) (Mackerras & Mackerras, 1944). Another predisposing factor for strike, especially in winter rainfall areas in Australia, is mycotic dermatitis, a skin infection caused by a mite, *Dermatophilus congolensis* (Murray & Wilkinson, 1980). Blood and pus on the fleece and skin are also strong egg-laying stimulants (Watts & Merritt, 1981).

The chemical identity of the attractive volatiles released by wool, blood, urine and so forth is not really well known. A great deal of work has been carried out on this subject, but the general picture remains confused. Sometimes the results from one study conflict with those of another, probably because the conditions in the various experiments were not identical. Flies of the same species may behave differently in different experiments because of differences in their physiological condition or age, or simply because they come from a different populations whose members behave differently. For these reasons it is frequently difficult to compare results meaningfully. For example, a number of sulphur compounds, like ethyl mercaptan, dimethyl disulphide and hydrogen sulphide, have been found to attract blowflies in some experiments. *Lucilia cuprina* was found to be attracted to ethyl mercaptan in one series of experiments (Freney, 1937), but not in others (Hepburn & Nolte, 1943). Freney also found that meat became more attractive when treated with calcium carbonate, a substance which increases the amount of volatile sulphur compounds given off from the meat. Further work showed that *Lucilia sericata* would lay eggs on moist clipped fleece treated with ammonium carbonate and indole, a compound that occurs naturally in dung. This species also laid eggs on sheep treated with ethyl mercaptan, hydrogen sulphide and ammonium carbonate (Cragg & Ramage, 1945). However, mercaptan-ammonium carbonate mixtures did not always elicit egg laying (Cragg, 1950*a*). Under natural conditions the sulphur compounds may be produced by the breakdown of the amino acid cystine in the keratin of the fleece, but experiments using cystine on living sheep failed to give positive results with *Lucilia cuprina,* perhaps because the cystine did not break down as expected (Cragg, 1950*a*). On the other hand, *Lucilia cuprina* was found to be more attracted to meat treated with cystine than to untreated meat (Hepburn & Nolte, 1943). A mixture of ethyl mercaptan, dimethyl disulphide and hydrogen sulphide was found to be a powerful attractant for *Lucilia sericata, L. caesar* and *L. illustris,* but for egg laying to occur, ammonia and carbon dioxide or indole had to be present (Cragg & Thurston, 1950).

Climatic factors, especially humidity and temperature, influence the olfactory behaviour of blowflies. Traps baited with ethyl mercaptan and hydrogen sulphide attracted *Calliphora vomitoria* strongly in October, but not at all in August (Cragg & Thurston, 1950); it is not clear whether this is a response to temperature or humidity, or to some other seasonal change.

We have seen that blowflies are attracted to urine-soiled fleece and that this may be due to the higher humidity. Some of the most important sheep-strike flies, like *Lucilia sericata,* have eggs that desiccate easily at 37° C (approximately the body temperature of sheep); this may explain why most strikes occur in the humid rump region (Davies & Hobson, 1935). Sheep sweat very heavily in the region of the shoulders, and strikes are also common there. Since most egg laying occurs during the morning when fly activity is at its highest, eggs will be hatching when dew is forming and when the sheep are lying on the wet grass (Davies, 1948).

Air and host temperatures are also important factors (MacLeod, 1949); both low and very high temperatures inhibit blowfly egg-laying activity. Species of *Lucilia* are generally active at higher temperatures than are species of *Calliphora,* and *Lucilia sericata* laid more eggs when substances known to induce egg-laying were exposed at 30 to 40° C than at lower temperatures (Cragg, 1956); the lowest temperature at which this species will lay eggs is about 14° C (MacLeod, 1947). The larvae of sun-loving species of *Lucilia* can withstand host tissue temperatures of 37° C, but *Calliphora* species cannot; even 32° C is fatal to them (Ratcliffe, 1935). Therefore, it is not surprising that *Calliphora* species will usually only attack animals whose body temperature is low. For example, myiasis by *Calliphora vicina* in woodmice and hedgehogs is correlated with torpor and low body temperature in these animals (Erzinçlioğlu & Davies, 1984). This probably explains why *Calliphora* myiasis in these small mammals occurs mainly in autumn when they are preparing for hibernation. The body temperature of such animals may drop considerably at such times; the skin temperature of hedgehogs will drop to that of the surroundings when the latter drops down to a critical level. The skin temperature of shorn sheep may fall to 26° C, and this may account for the occasional cases of *Calliphora* myiasis in such animals (Erzinçlioğlu & Davies, 1984).

The tropical blowfly *Chrysomya bezziana,* an obligate parasite, lays eggs on its host (usually a cow) about 2–3 hours before dusk. This is probably because the egg masses are known to suffer high mortality if experimentally exposed to solar radiation; there is thus a strong selection pressure on the flies to lay eggs late in the day. The high mortality is not due to dehydration, since the hatching rate of the eggs was reduced even further when they were kept moist (Spradbery, 1979).

Sheep tissues infested with larvae of *Lucilia cuprina* appears to be slightly alkaline (pH between 8 and 9), and experiments have shown that this is the optimum pH for the development of this blowfly species (Guerrini and others, 1988). It would be interesting to investigate the role of pH in the development of other species, and especially to see whether there are species that prefer a medium with a lower (more acidic) pH.

Very little is known about myiasis in small mammals in Britain, and observations on this would be of great interest. Rats, although seldom seen, are still extremely common and widespread in Britain, but no records exist of myiasis in rats in this country. Rats are known to inflict wounds on one another, and they frequent places that are attractive to blowflies, so one would expect wound myiasis to occur. The absence of records of this is probably due to the difficulties in observation. Dodge (1952) reports on a case of myiasis in a brown rat shot in the USA; apart from the bullet hole in its head, there were no wounds on its body. Clusters of fly eggs were found matted in its fur, and two species of *Lucilia* were eventually reared from it, including one, *L. illustris,* which occurs also in Britain. Clearly, blowflies will oviposit on unwounded rats. In view of the ecological importance of rats, studies of rat myiasis would be of particular interest. (Any studies on rats should be carried out with extreme caution. Rats are dangerous animals, not only because of the ferocity of their bites, but also because they harbour diseases that can be contracted by human beings.) There is also very little information on myiasis in other small mammals, such as mice, weasels, stoats and mink in Britain.

Why are some individual animals more susceptible to blowfly attack than others? Perhaps flies attack weak or inactive hosts. As we have seen, torpor and low body temperature may induce attack by *C. vicina,* and *Lucilia* species have been observed to select for attack individual sheep that are weak or diseased. The parasitized rat mentioned above was shot on premises that had been treated with rat poison four weeks earlier; it may have been poisoned, and therefore more susceptible to fly attack. Most patients suffering from human myiasis by *C. vicina* in Britain were probably both elderly and unwell at the time of infection.

Vertebrates other than mammals are also known to be victims of true myiasis, although cases of parasitized birds and reptiles are extremely rare. However, I have known two cases of myiasis in pet tortoises, *Testudo hermanni,* and the scarcity of records may be due to the difficulty of observing reptiles in the wild. Interestingly, a number of fleshflies (Sarcophagidae) are known to attack reptiles in countries other than Britain. Birds are the hosts of obligate parasites of the genus *Trypocalliphora* on the continent of Europe, but these species are not found in Britain.

Lucilia bufonivora is an obligate parasite of toads (*Bufo vulgaris*) in Britain. In Europe it is known to attack frogs and salamanders as well. The eggs are usually laid on the back or shoulders of the toad, and hatching is timed to coincide with the moulting of the toad, when there is a liquid exudate on the skin. The larvae invade the eyes and nostrils, killing the host and completing their development on the remains. This species of blowfly does not seem to be attracted to mammalian carcases for feeding; it visits only the carcases of amphibians. The larvae have particularly large mouth-hooks, giving the maggot the appearance of a walrus. Since the host dies soon after the attack and blowfly larvae continue their development on the dead host, it is probable that these flies oviposit on dead toads; this is a point worth investigation.

In cases of myiasis parasitism the blowfly larvae feed on flesh by releasing enzymes and sucking up the products of external digestion. Another kind of parasitism is blood-sucking by larvae. The only British species known to do this is *Protocalliphora azurea,* a parasite of passerine birds. The larvae of this species are thought to suck the blood of the nestlings. However, during a study of these flies, I examined a large number of swallow, starling and blackbird nests, and none of the nestlings showed any sign of feeding damage on the skin; furthermore, all such birds developed and fledged normally, even in a nest containing over 400 larvae. Perhaps, therefore, the larvae attack the parent birds, with their highly vascularised brood-pouches, rather than the nestlings. During the day, the blowfly larvae usually lie deep in the the nesting material, far from the nestlings; perhaps they come to the top of the nest material to feed at night when the adult birds are in the nest.

The larvae of *Protocalliphora* pupate in the nest material, and the adult flies never emerge until a day or so after the last bird has left the nest. This is presumably an adaptation to avoid predation; the trigger for emergence may be the sudden drop in temperature of the nest when the birds depart. Much remains to be discovered about the natural history of this very interesting blowfly/bird association. What do the larvae feed on, and when do they feed? Exactly where in the nest does the adult blowfly lay her eggs? Why are the blowfly larvae not attacked by the parent or nestling birds? How does infestation affect the birds?

Some blowflies parasitize invertebrates. The biology of *Bellardia cognata* is relatively well known (Keilin, 1919). This fly parasitizes the banded snail, *Helicella virgata,* laying its eggs near the opening of the mantle cavity. Only one larva appears to reach maturity on each snail. It attacks the organs of the host in turn. The snail dies before the larva is fully grown and the larva then completes its development on the host's remains. The fly is remarkably specific in its choice of host, avoiding closely-related species, and even avoiding a white non-banded variety of *H. virgata* in some localities but not in others. Much less is known about several other blowfly species, such as *Eggisops pecchiolii,* which also parasitize snails.

The familiar cluster fly, *Pollenia rudis,* is a parasite of earthworms. The fly does not attack its host directly, but lays its eggs on the surface of the soil; it is the first instar larvae that seek out and attack the worm (Thomson & Davies, 1973*a*). The larvae seem to descend into the soil only through natural pores, and are induced to penetrate the host by the presence of slime and coelomic fluid. The host, usually a species of *Allolobophora* or *Eisenia,* can autotomise parasitized segments, isolating and shedding the part of the body containing the blowfly larvae. The fly attacks many species of worm in the laboratory, but only a few in nature. High humidity is an important stimulus for egg-laying and the probability of attack may depend more on the habitat of the earthworm than on its species. Stream-bank species of worm are the most commonly attacked, while attacks are rare on species living in drier soils with a sparse vegetation cover and low surface humidities (Thomson & Davies, 1973*b*, 1974).

Thomson & Davies (1973*a*) list a number of aspects of the biology of *P. rudis* which remain unresolved. Is the presence of earthworms in the soil necessary for egg-laying? Whereabouts on the worm do the newly-hatched larvae penetrate? At what time of the year do the blowflies mate and lay eggs? Do the larvae, as well as the adults, overwinter? Almost nothing is known of the six other described species of British *Pollenia.* Are they earthworm parasites, or do they parasitize other organisms? An African *Pollenia* species is a honeybee parasite (Ibrahim, 1984).

Few blowfly larvae are predators. Certain tropical species, such as the African *Chrysomya albiceps,* are known to prey upon other fly larvae. No predatory species are known in Britain except *Stomorhina lunata,* a predator on the egg pods of locusts, which is found here only very occasionally. In years when there has been a great locust migration in Africa, some individuals are thought to be borne northwards by the wind. *S. lunata* has not yet been shown to attack any British grasshopper species, but this is not impossible, since at least four African locust species have been recorded as hosts. The biology of *S. lunata* in Africa has been studied by Greathead (1962).

No British blowflies are known to be predatory as adults, although some tropical species, such as species of *Bengalia,* prey upon other insects. No adult blowflies are known to suck blood, in the manner of mosquitoes and other biting flies, anywhere in the world, but some species will feed at wounds.

2.13 Plant associations

We have seen that nectar is a valuable energy source for blowflies (section 2.7) The flower-visiting habits of blowflies have some significance for pollination. Increasing attention is being paid in Britain to blowflies as pollinators of crops, especially in enclosed situations such as glasshouses and the pollination cages used by plant breeders (Free, 1993). There can be little doubt that they are of some ecological importance in this regard (Grinfel'd, 1955) and Pryor & Boden (1962) have demonstrated the importance of blowflies in the pollination of Eucalyptus trees in Australia. Flowers in the daisy family, Compositae, are said to be particularly favoured by blowflies, although the latter appear to be fairly catholic in their tastes. The antennae of *C. vicina* bear a number of flower odour receptors (Kaib,1974); this species is frequently seen on flowers, and an investigation of its role in pollination would be worthwhile.

A large number of plant species in various families possess flowers which mimic the appearance, temperature and odour of carrion, thus attracting blowflies and other carrion flies which pollinate them (Dafni, 1984). Some, such as Arum lilies, have been shown to produce volatiles that are known to attract blowflies (see pp. 19–20). One such plant commonly kept as a house plant in Britain is the succulent *Stapelia,* a native of Africa and Arabia. Hepburn (1943) has shown that its flowers attract females of *Lucilia cuprina* and induce them to oviposit. He also found that distillates from these flowers attracted the flies, although he did not identify the active compounds. The responses of British blowfly species to this tropical plant would be interesting to observe. I have seen females of *Calliphora vicina* settling and ovipositing on *Stapelia* flowers in Britain; the resulting eggs were observed to hatch, but the first-instar larvae failed to develop further. This may have been because the flowers began to dry out, or perhaps the larvae could not obtain nourishment from the flower; this question remains to be answered. However, some other carrion-fly-attracting flowers can support the complete development of larvae (Dafni, 1984). The *Stapelia* flower is said to smell like carrion, although it is not at all readily detectable by the human nose, except when the plant is placed in strong sunlight, when a slight odour can be detected. This odour suggests amines, such as trimethylamine, but this remains to be demonstrated. It would be interesting to see whether the active compounds are identical with the attractive fraction from decomposing carcases. In addition to their odour, *Stapelia* flowers have the dull reddish-brown colour of decomposing flesh and they generate a certain amount of warmth, mimicking the heat of decomposition; it is thought that these features enhance their attractiveness to the flies, although this has not been experimentally demonstrated. There can be little doubt that blowflies are important pollinators of *Stapelia.* The enormous flowers of *Rafflesia,* a parasitic plant from south-east Asia, also smell of carrion and are pollinated by the swarms of blowflies that are attracted to it.

The stink horn fungus, *Phallus impudicus,* smells of rotting meat and attracts large numbers of blowflies, as well as other kinds of carrion flies, which help to disperse the fungal spores. The flies associated with the stinkhorn in Britain have been studied by Smith (1956), who found that the relative numbers of *Calliphora* and *Lucilia* attracted to the fungus differed according to whether it was standing in the light or the shade, the *Lucilia* species dominating in the light. Some tropical blowflies, such as some African *Tricyclea* species, are known to be specifically fungus-breeders, but no such habit is known among the British species. Some British species whose habits are as yet unknown (such as species of *Pollenia*) may turn out to be fungus breeders.

2.14 Distribution and habitat preferences

Many of the more successful blowfly species, such as *Calliphora vicina,* are more or less uniformly distributed throughout the British Isles. On the other hand, some species show a distinctly regional distribution. *Cynomya mortuorum* and *Phormia terraenovae* are predominantly northern species, occurring mainly in Scotland and northern England, and only rarely in the south. *Calliphora uralensis* is confined, in the British Isles, to the northern and western parts of Scotland and western Ireland. Some species, like *Lucilia richardsi,* appear to be predominantly southern, being rare in Scotland and absent from the northwestern Highlands and Islands. Notes on the distribution of the various species are given in Chapter 5.

Blowflies also exhibit habitat preferences within their general area of distribution. For example, in northern Finland, Hedström & Nuorteva (1971) found that, from the birch forests at the base of a hill to its bare top, there was a gradual decrease in the numbers of flies of most families. The exception was the blowflies, which reached their greatest abundance in the sparse birch scrub near the timberline. The arctic blowfly *Boreellus atriceps,* which does not occur in Britain, was confined to the uppermost zone. The composition of the blowfly fauna of a Finnish city, together with the refuse depots around it, differed dramatically from the fauna of the nearby forests (Nuorteva & Laurikainen, 1964). *Calliphora uralensis* and *Phormia terraenovae* dominated in the urban areas, while *Calliphora vomitoria, C. loewi, C. subalpina* and *Cynomya mortuorum* dominated in the forests.

Few detailed studies of this kind have been carried out in Britain. MacLeod & Donnelly (1957) found that *Calliphora vomitoria* and *Lucilia ampullacea* apparently need dense cover such as woodland. *Calliphora vicina* and *Lucilia illustris* were more numerous in more open conditions, while *Lucilia caesar* preferred scrub with sparse trees. In France,

Holdaway (1930) found that *L. sericata* frequented open habitats, while *L. caesar* preferred shady habitats. In 1956, Macleod & Donnelly showed that *Phormia terraenovae* exhibited no preference between open or sheltered habitats within its range, and that *Cynomya mortuorum* was a fly of exposed habitats. This last finding was confirmed by Nuorteva (1972) in Finland.

Laboratory studies may give clues about the possible habitat preferences of different species. Thus Evans (1936) showed that high temperatures caused higher mortality in the pre-pupae of *Phormia terraenovae* than in those of *Calliphora vicina*. Another interesting finding was that high humidities shortened the life of *Ph. terraenovae* adults. MacLeod & Donnelly (1957) were the first to suggest that male and female blowflies may have slightly different habitat preferences. Using baited traps, their catch showed a higher male:female ratio of *Calliphora* and *Lucilia* species in canopy habitats than in open habitats, although this was not confirmed by unbaited, tent-trap catches. Nuorteva (1959) found in Finland that the proportion of males was highest in the samples of blowflies trapped in the least favourable areas. However, Norris (1959) in Australia showed a strong positive correlation between the proportion of males and the total number of flies caught, suggesting that males tended to leave favoured habitats less than females. The relationship between habitat and sex ratio remains unresolved and requires further investigation.

Little is known of the activity of blowflies at various heights above the ground in a given locality. Roberts (1933), in Texas, attempted such a study, but caught so few specimens that he could draw no meaningful conclusions. He did, however, show that blowflies could be trapped up to 13.5 m above the ground. A weakness of this study was that meat-baited traps were used; these would undoubtedly have attracted the flies and thus the results cannot be interpreted as indicating that blowflies normally occur at such heights. Williams (1954), in New York, found that traps placed on top of buildings at heights of 25.5 m caught significantly fewer blowflies than did those at ground level. It would be interesting to set up traps at ground level and at canopy level in a woodland to see whether species differ in their vertical distribution.

The habitat preferences of blowflies are not well known, and deserve further attention. Furthermore, much of what has been published on the subject may be misleading, since observed differences in habitat selection may be due to other factors, as the following section will show.

2.15 Activity patterns

It is a matter of common experience that flies are not active all hours of the day or all days of the year. In spite of this, or possibly because of it, very few studies have been carried out on the seasonal and, especially, the diurnal activity patterns of blowflies, and most such studies have been carried out in countries other than Britain.

Diurnal activity patterns are usually either unimodal, with one peak of activity per day, or bimodal, with two daily peaks of activity. Insects in which activity increases with light intensity and temperature often show a unimodal pattern with the peak at the warmest time, around midday; but if the weather gets too hot at midday activity may be suppressed at that time, producing a bimodal pattern. In Central Asia, the diurnal activity curves of *Calliphora vicina* were unimodal during the cooler months, but strongly bimodal in the hotter months of the year (Sychevskaya, 1962). During the hotter months activity was at its lowest at midday when temperature and light intensity were at their highest; at such times the flies sheltered in dark places such as cellars. Similar results were obtained in an Australian study (Norris, 1966). In the cool climate of Finland in August, *Calliphora vomitoria, Lucilia caesar* and *L. illustris* exhibited a unimodal activity curve, with its peak at midday (Nuorteva, 1959).

In sub-arctic Finland in July, blowflies exhibited a unimodal pattern with a peak just after noon, but all other flies studied exhibited a bimodal pattern with a midday minimum (Nuorteva, 1965). Some species (*Calliphora alpina* and the non-British *Lucilia fuscipalpis*) were active only in the afternoon, whereas others, such as *Cynomya mortuorum* and *Calliphora uralensis,* were often active as early as 6 a.m., the latter even at temperatures of less than 10° C. A similar result was obtained in the Kevo River Nature Reserve in northern Finland (Hanski & Nuorteva, 1975). Here, too, blowflies showed unimodal activity patterns throughout the area; most other flies exhibited a bimodal pattern on the fells, but a unimodal one in the river canyon. This persistent unimodality of blowflies (in contrast to the bimodality of other flies) in sub-arctic areas, where the temperature is seldom high enough for insect activity, rather suggests that acclimatisation to low temperatures has reduced the tolerance of non-blowflies to the heat of insolation, whereas blowflies remain tolerant to both extremes. The metallic colouring of most blowflies will reflect the sun's radiation and help protect them from overheating, although because of their large size they are not expected to lose heat as quickly as smaller flies (Willmer & Unwin, 1981). It is intriguing, therefore, that smaller, but equally metallic, flies were not active during the warmest part of the day in Hanski & Nuorteva's study, since they might have been expected to be able to cope with the problem of overheating better than the larger blowflies.

Digby (1958*a*, *b*) showed in the laboratory how flight activity in *Calliphora vicina* depended on light intensity, temperature and wind speed. A wind speed of 0.7m sec⁻¹ was optimal; higher speeds inhibited flight.

Barometric pressure is an environmental factor that seems to have received little attention in ecological studies. Edwards (1961) showed that falling pressure, from an initial state of rising or level pressure, resulted in an increase in the flight activity of *Calliphora vicina* (but not *C. vomitoria*); it was the change to falling pressure, not the falling pressure in itself, that caused the change in behaviour. Interestingly, rising pressure, from an initial state of falling or level pressure, had little or no effect on flight activity.

Smith (1983) presented evidence that spontaneous flight activity in the Australian blowfly *Lucilia cuprina* was controlled by an internal 'clock' with a period of roughly 24 hours. Flies maintained their diurnal activity patterns even when kept in constant light or constant darkness, and at constant temperature. Other species may exhibit similar behaviour.

There is great scope for original work on the diurnal activity patterns of blowflies in Britain. It would be particularly interesting to compare the periodicity of the widespread and ecologically important species *Calliphora vicina* in southern and northern parts of the country. Other species of interest in this regard include the relatively little-known cold-adapted species, *Calliphora alpina, C. uralensis, Cynomya mortuorum* and *Phormia terraenovae*. It is essential in such studies to sample at regular intervals (say hourly) throughout the 24-hour period; otherwise, as Norris (1965) points out, one might miss significant periods of activity.

2.16 Population densities and population changes

It has proved surprisingly difficult to estimate the population density of blowflies. The commonest method is fraught with difficulties. It involves releasing a known number of marked flies, letting them mingle with the rest of the population, and then trapping a large number of specimens; the percentage of marked flies in the total catch will give an indication of the population size. The usual index used in such studies is the Lincoln Index, which is calculated as follows:

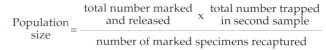

$$\text{Population size} = \frac{\text{total number marked and released} \times \text{total number trapped in second sample}}{\text{number of marked specimens recaptured}}$$

There are several problems with this method.

1. It is assumed that the capture and marking of the insects will not harm them in any way, and will not affect their behaviour after release.

2. It is very difficult to define satisfactorily the spatial limits of the area in which the fly population is to be estimated.
3. It is assumed that there will be no changes in the population during the period between release and recapture. This is not always true. For example, some flies may die, while others may emerge; or some individuals may leave the population (emigrate), while others may enter it (immigrate). For this reason it is important not to have too great a delay between release and recapture.
4. It is assumed that the released flies will mingle with the population as a whole after release, and not remain together as a distinct sub-group. It is necessary, therefore, to leave enough time for the flies to distribute themselves among the population before recapture. It is clear that a balance must be struck; the period elapsing before recapture must not be too long for reason (3) above, and not too short for reason (4).
5. Flies must be recaptured under conditions as close as possible to those under which they were initially captured. This is because the flies may behave differently or distribute themselves differently under different conditions.
6. The second catch must include at least 10% of the marked individuals to give results of value.

In spite of these difficulties, the Lincoln Index may be useful as long as the results of any piece of work are interpreted with these limitations in mind. In studies of blowflies, it is particularly useful, not so much as an indicator of absolute population size, but as a method of comparing the density of different species in a particular locality, or of a particular species at different times of the year. Chalmers & Parker (1989) discuss release-recapture methods in more detail.

1 acre = 0.4 ha

Using this method, MacLeod & Donnelly (1957) estimated the population density of *Calliphora vicina* in an area near Carlisle to be 50 to 200 flies per acre in August, 400 to 1000 in September and 700 to 1000 in October. Flies of the *Lucilia caesar* group (*L. caesar*, *L. illustris* and *L. ampullacea* combined) were estimated at 40 to 70 per acre in August and September, but less than one individual of *L. caesar* per acre in October. The figures are interesting as an indication of the relative abundance of the different species at different times of the year. In a later year and in another area, near the Solway estuary, the actual population density estimates were much lower, but the same pattern of seasonal abundance was observed for *C. vicina* (MacLeod & Donnelly, 1960). On the Roxburgh-Dumfries border, *Cynomya mortuorum* increased markedly in abundance in September (MacLeod & Donnelly, 1958).

A complicating factor in population density estimates is the phenomenon of micro-geographic aggregations. MacLeod & Donnelly (1962) found local aggregations of blowflies in nature. These were often associated with distinct habitat units, such as vegetation patches or transient food

sources like animal carcases, but they were also observed in
areas that were apparently uniform ecologically. Many were
found to persist over a series of occasions, and to be resistant
to disruption. What causes these aggregations? Are they due
to local variations in physical factors, such as humidity or
other aspects of microclimate?

An interesting study on blowfly abundance was
carried out on the islands of a large Finnish lake by Nuorteva
& Räsänen (1968), who found that large islands supported
more blowflies per unit area than did smaller islands. Most
blowfly species exhibited a maximum abundance later in the
year on the islands than on the nearby mainland; this was
thought to be due to the cooling effect of the slowly warming
water mass around the islands. In June, blowflies were not
detected on the smallest islands, although they had already
emerged on the mainland and on the larger islands. Little is
known of the species composition of the blowfly faunas of
British off-shore islands. Do certain species frequent large
islands, others small ones? Why? Do blowflies overwinter on
these islands, or do they re-colonise annually?

Why do populations sometimes fluctuate, and
sometimes remain constant? Food availability may be
important. An interesting laboratory study on the Australian
species *Lucilia cuprina* by Nicholson (1954, 1957) explored
this possibility. When the blowflies were kept under
conditions that forced the larvae to compete strongly for
food, the populations exhibited strong periodic fluctuations.
If only 50 g of liver were provided daily for the larvae (while
the adults received plenty of food) the numbers of adults
fluctuated over several generations from a maximum of 4000
to a minimum of none, when all the surviving individuals in
the population were larvae or eggs.

The fluctuations in these experiments were explained
in the following way. When many adults were present in the
population, many eggs were laid. This resulted in strong
larval competition for the limited food supply. Hardly any of
the larvae from eggs laid during adult population peaks
survived, mainly because they did not grow large enough to
pupate. Large adult populations, therefore, gave rise to few
adults in the next generation, the population declined and
very few eggs were laid. This resulted in the survival of
larger numbers of larvae; these were able to pupate, and the
adult population increased again. When the adults, as well as
the larvae, were given a restricted amount of liver, fewer
eggs were laid because the adults, as we have seen, require
adequate protein to enable them to mature their eggs. Under
these conditions the population remained more or less
constant, and did not fluctuate.

Two interesting conclusions emerge from these
results. The first is that the harmful effects of competition at
high densities (density-dependent effects) are of two kinds:
those that affect oviposition immediately (as when both
adults and larvae were given a limited food supply), and
those that affect oviposition only in later generations (as
when only the larvae were given a limited diet). The second

conclusion is that only those density-dependent responses that exhibit a time lag result in observed fluctuations in populations. Gurney and others (1980) have since confirmed Nicholson's conclusions.

Population changes can occur for other reasons. Thus MacLeod & Donnelly (1958), working in the Southern Uplands of Scotland during August and September 1953, showed that populations of *Lucilia caesar, Calliphora vicina* and *Calliphora vomitoria* shifted seasonally from the more sheltered habitats to the more exposed. This phenomenon has not been adequately explained, but may possibly be due to a greater availability of sunshine in the more open habitats. The factors affecting population shifts of this kind are in need of further investigation. In this connection, two studies in the United States are worth mentioning. In the first, Schoof & Savage (1955), working in urban environments, showed that the population curves of several blowfly species exhibited little congruence between the years 1949 and 1950. They suggested that the cooler conditions and excessive rainfall in 1950 may have been responsible for the lower fly abundance in that year. However, Dicke & Eastwood (1952) in a study which was similar, but was carried out in a more rural environment, showed that there was substantial agreement between the successive annual population curves for the years 1949 and 1950, even though their study area also experienced cooler, wetter weather in 1950. Stewart & Roessler (1942) failed to demonstrate any close correlation between blowfly numbers and temperature and saturation deficits. Perhaps the dramatic changes in blowfly populations from year to year that occur in urban areas (as opposed to rural habitats) are due to changes in standards of hygiene and rubbish disposal practices; the presence of refuse depots can influence the abundance and species composition of blowflies in cities (Nuorteva & Laurikainen, 1964).

2.17 Dispersal and migration

Blowflies can fly for long periods and cover great distances (section 2.2). Natural features such as rivers and woods do not appear to be obstacles to blowfly dispersal. *Calliphora vicina* and flies of the *Lucilia caesar* group marked on one side of a 200-metre wide river were recaptured on the farther side, and marked *Lucilia caesar, L. sericata* and *Calliphora vicina* released near a deciduous wood belt 90 metres wide were recovered beyond the wood (MacLeod & Donnelly, 1960). *Calliphora* species, *Lucilia sericata* and *Phormia terraenovae* were able to disperse up an incline of approximately 45 degrees extending to a vertical height of about 167 metres above the release point (MacLeod & Donnelly, 1958). Hightower & Alley (1963) presented evidence that the American species *Cochliomyia hominivorax* was channelled to traps by terrain features, although MacLeod & Donnelly (1958) found no evidence of canalisation of dispersal by ground contour in a number of

British species. Schoof & Mail (1953), working in North America, found some irregularity in the dispersal of marked-released *Phormia regina*; most specimens were recovered from hilly areas, perhaps indicating that this cold-adapted species reacted to temperature gradients. It has been suggested that the prevailing winds may affect the direction of blowfly dispersal (Gurney & Woodhill, 1926), but this has not been adequately demonstrated.

MacLeod & Donnelly (1963) suggested that two types of flight are recognisable in flies dispersing from an initial concentration. The first type, called the exodus flight, is unidirectional, and either fast or sustained, covering considerable distances quickly, at speeds of over 11 km h^{-1}. The second type, called the random-direction flight, involves little or no net displacement, and the direction of each new short flight is independent of the direction of the preceding flight. Doubt has been cast (Norris, 1965) on MacLeod & Donnelly's idea that there would be a clear-cut distribution of any batch of flies into two such groups. An investigation of this phenomenon would be desirable.

How do blowflies select their direction of dispersal? Do they fly towards certain features of the landscape? Do they follow temperature or humidity gradients, or, as suggested above, do they follow the direction of the prevailing winds? Shura-Bura and others (1958) showed that *Phormia terraenovae* released in the countryside moved towards towns. There is ample scope for research.

Experiments of this kind must be carefully designed. Laboratory-bred flies released in the field may not behave naturally. Little information is available on the dispersal of flies emerging from their puparia in nature.

Little is known about the possible large-scale migrations of blowflies. Williams and others (1956) recorded southward movement of *Calliphora vicina* and *C. vomitoria* in autumn. These and other species may be migratory in at least part of their range; more research is needed.

2.18 Competition in carrion

Blowflies are not the only organisms that feed upon carrion. A large number of species of insects, other invertebrates, bacteria and fungi will colonise carrion under the right conditions. Although blowflies are usually the earliest insect colonisers of carrion during the summer, vertebrate scavengers, such as dogs, foxes and crows, will often feed upon a carcase during the earlier stages of decomposition. The role of such scavengers in Britain is not important, unlike the situation in such places as the savannahs of Africa or the the northern Tundra, where there are large numbers of scavenging vertebrates. In certain parts of Africa, scavengers may remove up to 89% of the available carcases (Richardson, 1980).

As with most other natural resources, there exists a keen competition for carrion in nature. Two kinds of competition can be identified; competition between individuals of the same species (intraspecific competition) and competition between different species (interspecific competition). Intraspecific competition on a carcase should, in theory, result in high mortality, but observations on carcases in nature have shown that such competition results in the production of smaller-sized flies that lay fewer eggs, and that mortality is low (about 15%). This sort of statement, however, is usually made on the basis of anecdotal evidence and further observations are needed. It is interesting to note that, under laboratory conditions, great mortality may occur at the egg stage, and this may be overlooked during studies in the field.

Interspecific competition results when two or more species occur on the same carcase. It is often stated that different species having identical niche requirements cannot coexist. If one accepts this contention it follows that maggots of different species inhabiting the same carcase are likely to be doing different things within the carcase. Various studies have shown this to be broadly true. Fleshfly (Sarcophagidae) larvae are often found on the same carcases as blowfly larvae, although they are believed to be less efficient than blowflies in exploiting carrion. Sarcophagids have been shown to avoid competition with blowflies in two ways. First, many sarcophagid species give birth to live larvae, rather than eggs; being at a more advanced stage of development when laid on the carrion than blowfly larvae, which are deposited as eggs, the sarcophagids have a 'head start' and can complete their development before the blowfly larvae reach the third instar when they are competitively far superior to the fleshfly larvae (Denno & Cothran, 1975, 1976). Another, very unusual, way that sarcophagids have been shown to deal with competition from calliphorids is by killing them. In a study by Blackith & Blackith (1984), it was shown that first-instar larvae of *Sarcophaga aratrix* could attack and kill third-instar larvae of *Calliphora vicina.* Interestingly, *S. aratrix* would also attack and kill another species of *Sarcophaga, S. subvicina,* although the latter species did not attack either *S. aratrix* or *C. vicina.*

We have seen above (section 2.11) that many insects and other organisms may feed upon blowfly larvae. Although this phenomenon of predation is distinct from competition, it is difficult to draw a hard and fast line between the two phenomena when considering normally carrion-feeding larvae preying upon other such larvae. Species of *Muscina* (Muscidae) are known to prey as larvae upon blowfly larvae in Britain, while the larvae of different species of blowfly are known to be predatory on one another in other parts of the world (for example, the African *Chrysomya albiceps* preys upon *Chrysomya regalis*). In such cases, the larva is clearly a predator but, since it is also destroying another organism feeding on its other food source, the carrion itself, it is also a competitor. It would be interesting to see whether any British blowfly larvae prey upon other blowfly larvae, and to discover further species of non-blowfly larvae behaving in this way. Studies of this sort would help us to understand what the different species of carrion flies are doing in the carcase.

In earlier sections we discussed the seasonality of blowflies and their habitat and carcase preferences. It is clear that such differing preferences may contribute to the avoidance of competition.

It is now well established that *Calliphora vomitoria* is a coloniser of large carcases in nature and tends to avoid small carcases and so appears to avoid competition with *Calliphora vicina* (Blackith & Blackith, 1990; Davies, 1990; Erzinçlioğlu, 1986a). The larvae of *C. vomitoria* develop more slowly than those of its close relative, *C. vicina,* and it would appear to avoid competition with the latter in small carcases. *C. vomitoria* is also the most abundant carrion blowfly in many areas of upland England and Wales, and may avoid competition with other species for this reason (Davies, 1990). *Lucilia* species have been found to be abundant up to an altitude of 300-430 m, but were rare or absent above 500 m, except during exceptionally hot weather (Davies, 1990).

It would be interesting and useful to investigate the problem of niche separation and competition in British blowflies. Few such investigations have been carried out anywhere in the world, but a study of seven abundant species on the Highveld of the Transvaal, South Africa, showed that they were separated by habitat preferences, seasonal occurrences, the chronological order in which they arrived at the carcases (the faunal succession) and differences in larval behaviour (Meskin, 1986). Another study well worth referring to is that of Hanski (1987). Putman (1983) provides a comprehensive account of carrion decomposition in nature.

Another question that is rarely addressed concerns the effects of various environmental conditions on competition. For example, if two species are feeding as larvae in a carcase, will one compete more successfully at higher temperatures and the other at lower temperatures? For further ideas on this subject see Wells & Greenberg (1992).

2.19 Fossil blowflies

The oldest alleged fossil blowfly puparia known are a few specimens from the Late Cretaceous (about 70 million years ago) of Alberta, Canada, which were described by McAlpine (1970) as *Cretaphormia fowleri.* Are these really the remains of blowflies? Having examined them myself, I conclude tentatively that they probably are the remains of blowflies or blowfly-like flies. If they are indeed blowflies, then this group of insects is much more ancient than was once thought, and would have been contemporaneous with the dinosaurs. The more conventional view, however, is that the blowflies arose sometime during the Tertiary, probably around 20–30 million years ago. The oldest undoubted fossil blowfly puparia are some remains found in association with *Australopithecus* bones in the Makapan Valley, South Africa, dating from the Tertiary/Quaternary boundary (about 1–2

PLATE 1

Blowflies

1. *Calliphora vicina*

2. *Lucilia sericata*

3. *Cynomya mortuorum*

4. *Phormia terraenovae*

5. *Protocalliphora azurea*

PLATE 2

Blowflies

1. *Melinda gentilis*

2. *Pollenia rudis*

3. *Pseudonesia puberula*

4. *Bellardia agilis*

5. *Eggisops pecchiolii*

PLATE 3

Heads, from the left

1. *Calliphora vicina*
2. *Calliphora vomitoria*
3. *Calliphora uralensis*
4. *Calliphora alpina*
5. *Lucilia sericata*
6. *Calliphora loewi*
7. *Phormia terraenovae*
8. *Cynomya mortuorum*

Puparia

9. *Calliphora vicina*
10. *Phormia terraenovae*

1

2

3

4

5

6

7

8

9

10

PLATE 4

Flies that resemble blowflies

1. *Sarcophaga carnaria*
   (Sarcophagidae)

2. *Gymnochaeta viridis*
   (Tachinidae)

3. *Mesembrina meridiana*
   (Muscidae)

4. *Muscina pabulorum*
   (Muscidae)

5. *Dasyphora cyanella*
   (Muscidae)

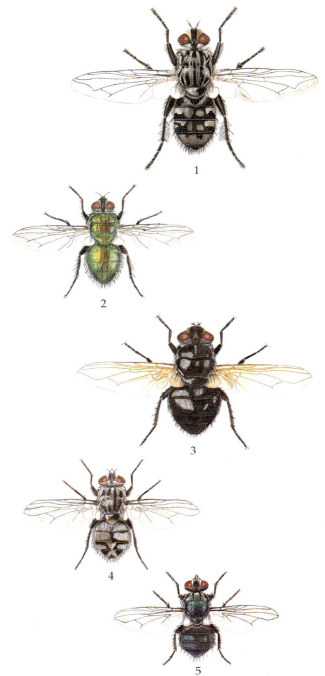

million years ago) (Kitching, 1980). The oldest fossils that
have been positively identified to species are some puparia
of *Phormia terraenovae* that were found among the remains of
woolly rhinoceros and steppe bison from the Late Pleistocene
(about 75,000 years ago) of Belgium (Gautier & Schumann,
1973; Gautier, 1975).

More recently, the remains of mammoths have been
discovered in Shropshire, associated with a large number of
Phormia terraenovae puparia (Coope & Lister, 1987). Some of
these puparia are illustrated in fig. 11. Although the puparia
are misshapen, various structures such as the rear spiracles,
the spines and the cephalopharyngeal skeleton have been
perfectly preserved. It is interesting to note that all
identifiable blowfly remains from European Ice Age
mammal bones proved to be *P. terraenovae*. The association of
these flies with low temperatures is well documented
(Erzinçlioğlu, 1988).

Puparia are very durable objects, whereas adult flies
are particularly fragile. It is likely, therefore, that many
fossilised puparia exist in nature but are overlooked, and it is
also likely that future palaeontological research on blowflies
will depend on finds of puparia. There are several areas of
Britain that would repay investigation in this field; the
interbasaltic Ardtun Leaf Beds (Eocene) of the Isle of Mull,
Scotland, seem particularly promising. These are about 60
million years old, and therefore a likely source for fossils of
early blowflies. Representatives of two other families of flies,
Tipulidae (craneflies) and Bibionidae (March flies), have been
discovered in these beds (Zeuner, 1941).

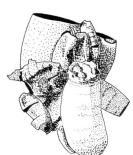

Fig. 11. Puparia of *Phormia
terraenovae*, found associated
with mammoth remains.

3 Blowflies and man

3.1 Introduction

We have already seen that various aspects of the biology of blowflies are relevant to human activities. In this chapter these aspects will be discussed in greater detail, and placed more firmly into context. In addition, some other ways in which blowflies can impinge on human activities will be explored.

3.2 Human associations

A number of blowfly species are frequently found in association with human dwellings, or the dwellings of domesticated animals. Flies that show this sort of association with man are termed 'synanthropic', and the phenomenon is known as 'synanthropy'. Some species seem to be synanthropic wherever they are found, while others behave in this way only in certain parts of the world, or in certain habitats within a particular geographical area. For example, *Calliphora uralensis* is synanthropic in parts of Scandinavia, but it is not known to be so in Britain; *Cynomya mortuorum*, not normally synanthropic in most of Europe, is found associated with human dwellings in the Alpine regions of Central Europe.

synanthropy: an association with the dwellings of man or domestic animals

The set of environmental conditions created by man must be attractive to a synanthropic fly species. This means that certain requirements of the species in question must be satisfied, such as food sources for both the larvae and adults, as well as acceptable physical conditions such as light, temperature and humidity. Attempts have been made to classify the different kinds of synanthropy, and the term 'eusynanthropy' has been proposed for species whose entire life cycle takes place within the man-made environment. Species which are synanthropic during only part of their life cycle are termed 'hemisynanthropic', while 'asynanthropic' flies are species not associated with man. The term 'symbovine' has been used to describe flies which are associated with the excreta of farm animals, whether in the field or in barns and stables. These terms are useful, but they should not be interpreted too rigidly since they essentially describe different points on a continuum.

Nuorteva (1965b) tried to classify blowflies along a synanthropic continuum using a Synanthropy Index. In order to collect data for the calculation of this index, flies are trapped in comparable ways and simultaneously at three sites: a city, an isolated rural house and in the wild. For a given species, the index is calculated as follows:

$$\frac{2a + b - 2c}{2}$$

where:

a = percentage of specimens trapped in a city relative to all specimens of that species collected at all three sites,

b = percentage of specimens trapped at the isolated rural house relative to all specimens of that species collected at all three sites, and

c = percentage of specimens collected in the wild relative to all specimens of that species collected at all three sites.

The index ranges between +100 and -100. A score of +100 indicates the greatest possible degree of synanthropy. Negative values indicate that the species avoids man.

Little is known about the synanthropic behaviour of many blowfly species in Britain. A species like *Calliphora vicina* is usually regarded as being synanthropic in most parts of its range, but little is known of the synanthropic behaviour of other species in upland or coastal regions. Is *Calliphora uralensis* synanthropic in the northern and western parts of Scotland? Are species like *Calliphora alpina* and *Calliphora subalpina* synanthropic in the upland areas of Britain? In the southern part of its British range, *Cynomya mortuorum* appears to be restricted to coastal areas; is it synanthropic in these areas? These and many other questions await investigation. A useful discussion of fly synanthropy is given by Povolny (in Greenberg, 1971).

3.3 Blowflies and disease

Flies, including blowflies, have often been implicated as the agents of mechanical transmission of disease. The idea is that the flies pick up bacteria from carcases and dung and then settle on food intended for human consumption, transferring part of their bacterial load onto the food. Other routes of infection include the faeces deposited by flies on food, as well as their 'vomit drops', the drops of fluid they regurgitate onto the food to soften it before feeding. Whether or not flies can actually transmit diseases in these ways has long been, and continues to be, a matter of dispute. The evidence is of three kinds. First, laboratory studies have shown that agar plates upon which wild-caught flies had been allowed to walk developed rich cultures of various disease-causing bacteria. Secondly, laboratory attempts to transmit diseases to animals using blowflies have often been successful. Thirdly, epidemiological studies have shown that a decline in fly numbers is often correlated with a decline in diseases thought to be fly-borne. This last line of evidence is usually considered weak since it is difficult to tell whether one is dealing with a 'cause-and-effect' situation or whether the disease and the flies declined together because of the effect of a third factor, such as air temperature, of which the investigator may have been unaware. A classic British study early this century showed that the decline of horse-drawn vehicles correlated with a decline in the death rate caused by summer diarrhoea, and it was concluded that the decrease in the amount of horse manure on the roads reduced the

numbers of fly-breeding sites resulting in fewer flies and
hence less diarrhoea. The fly species in this case was the
housefly, *Musca domestica*, which is not a blowfly, but the case
is a good example of this kind of study.

Although the case against flies is not proven there is
enough evidence to suggest that these insects play an
important role in disseminating summer enteric diseases in
Britain. Little work has been done on the species of blowfly
associated with dung, especially dog dung, in cities, and
there is scope for basic studies to find out which species are
attracted to dung in what parts of the country. British
blowflies, as we have seen in Chapter 2, do not breed in
dung but are frequently attracted to it obtain a protein meal.
In fact, it seems to be a general rule that the species most
commonly visiting dung as adults are not those which most
commonly breed in it. Some information on the fly fauna of
dog dung in British cities is given by Disney (1973) for the
City of Bath and by Erzinçlioğlu (1981) for London.
Greenberg (1971, 1973) gives an excellent introduction to the
problem of flies and disease.

3.4 Blowflies in war

One of the unfortunate but inevitable results of war is the
death or injury of large numbers of soldiers on the
battlefield. Such a concentration of dead bodies will, of
course, attract blowflies to breed in large numbers. The
medical significance of such situations is obvious and it has
long been realised that the control of flies during wartime is
essential if the troops are not to succumb to fly-borne
diseases. During the First World War armies on the Egyptian
and Mesopotamian fronts, in particular, adopted strict
measures to control flies by disposing of faeces, as well as by
burying the bodies of both men and horses with the
minimum of delay. Although such matters are of mainly
historical interest nowadays, because of the great advances in
medicine and hygiene made since the Second World War, it
would be a worthwhile exercise to compile a history of this
problem since early times. Maxwell-Lefroy (1916) and
Davidson (1918) give interesting insights into the magnitude
of the fly problem during the First World War.

Injured soldiers on the battlefield were open to
another danger: the danger of myiasis. As we have seen in
Chapter 2, many blowfly species will lay eggs on the open
wounds of animals and humans. This is particularly likely to
happen where flies are abundant, as they would be on a
battlefield. Again, during the First World War, the last great
war before the large-scale advent of modern medicine,
myiasis was a major problem in men and horses, particularly
in the trench warfare in France. The agony caused by such
infections could be very great, not only because of the
infection itself, but also because of the feelings of revulsion
engendered in the victim. The following extract from the
diary of a young subaltern in the 5th Inniskillings, wounded

on a beach in Gallipoli in 1916, gives some idea of the physical and psychological effects of myiasis:

"With my free hand I took off my puttees at my leisure and bound them round my head. Next came the ampule of iodine, which I broke and poured into my shoulder through the torn shirt. It seemed to attract the flies, who came, green- and blue-bottles, in dozens to the feast. I began to stink horribly in the sun...I heard a little rustle in the bush behind me. It was the water carrier. A real boy with real water came and knelt beside me, giving me drink and talking to me...After a while I sent him off because I was stinking so vilely, telling him to let someone know where I was in case the wounded could be moved that night. My shoulder was by this time full of maggots...Though there was no other wounded man in sight, the whole valley was resounding with that ghastly cry, 'Stretcher-bearers! Stretcher-bearers!' and awful curses...No stretcher-bearers came...I turned over and lay on my face in the sand".

This particular subaltern was eventually rescued, but he records that:

"the stench of wounds and the swarms of flies [were] quite undescribable...There was no shade - nothing but flies and stretchers".

In spite of the horrors of myiasis during wartime, it is interesting to recall that this very condition was put to good medical use during the First World War. The larvae of some blowfly species will feed on wounds, but will restrict their attentions to putrefying tissues and will not attack healthy ones. Furthermore, the bactericidal properties of the maggots encouraged the growth of healthy tissue. This discovery was made by Baron Larrey, one of Napoleon's surgeons, but it seems not to have been applied medically to any extent until the First World War. During that war, maggots were reared under sterile conditions and applied to the wounds of soldiers in the field, especially in cases of osteomyelitis (bone infection). The method was often successful, but it sometimes failed because, in certain cases, the maggots did not confine their attentions to the diseased tissues, but attacked healthy tissues as well. This happened because the medics did not distinguish between different species of blowfly, and sometimes they inadvertently included species which habitually attack healthy tissues. This was not understood at the time, and the method of treatment fell out of favour. Recently, there has been a renewed interest in this method, especially in the United States; a review of the history and medical significance of 'maggot therapy' is given by Sherman & Pechter (1988).

3.5 Forensic applications

A dead human body, if left exposed above the ground, will attract blowflies to lay eggs on it. Human bodies that are left exposed in this way are usually murder victims or sudden deaths, and thus are a matter for police investigation. The maggots that are often found on such bodies may contribute to a police investigation in a number of ways. First, an estimate of their age will give some idea of the minimum time since death; one can usually make a statement, based on the estimated age of the maggots, such as "death must have occurred not later than midnight on Wednesday". In other words, if we believe that the maggots were five days old, then death could not have occurred later than five days ago. It might have occurred much earlier than that; one may not know whether the body was exposed to blowfly activity immediately after death, or whether it was concealed in some way, only to be exposed at a later date. In cases where there are good reasons for supposing that the body was exposed immediately after death one may be able to estimate the date of death more precisely. It is also necessary to consider whether the temperatures were high enough for the flies to be active during the period before the blowfly eggs must have been laid. Even if a body was exposed immediately after death, the blowflies may have been inactive at that time, rendering the body effectively unavailable for blowfly oviposition.

As we have seen, temperature is the single most important factor affecting the rate of larval development. To estimate the age of the maggots, we need to know the temperature at which they were developing. In a murder case this information is not directly available, but the temperature of the larval mass can be measured when the body is discovered. Meteorological records may help, but these must be interpreted with caution, because a meteorologist is usually interested in the 'big picture' of the weather, and goes out of his way to avoid measuring the temperature near the ground - the microclimate - which is precisely what the investigator of a murder case is interested in. Meteorological data have to be interpreted, therefore, in conjunction with temperature measurements made at the scene of the crime.

We still need to know more about the rate of development of different species at different temperatures. Because temperatures in nature are never constant, it is particularly valuable to conduct experiments at both fluctuating and constant temperatures; this has not often been appreciated in the past. It has often been assumed that maggots developing at, say, 15° C will develop at the same rate at fluctuating temperatures of 10–20° C (mean = 15° C). My own (unpublished) work has shown that maggots of *Calliphora vicina* will develop more slowly in the fluctuating temperature regime. Greenberg (1991) has shown that the same is true of certain North American species. Further research is needed.

Blowflies can also help the police to discover the place where the victim died. In this context, 'place' may be used in an ecological or a geographical sense. A place in an ecological sense is a habitat, such as a woodland, a seashore, a house, or the top of a mountain. Geographically, a place is London, the north of Scotland or the island of Jersey. It is clear that a knowledge of blowfly distribution, habitat preferences and synanthropy may help us to answer the question "Where did death occur?", particularly in murder cases in which the body is infested with the larvae of a rare or localised species. Sometimes the absence of blowfly larvae from a body, when they would have been expected if the story told by the accused person had been true, can give useful clues as to where the body may have been. An example of such a case has recently been recorded (Erzinçlioğlu, 1989a). Since a body exposed above ground during the summer is almost certain to attract blowflies in large numbers, suspicion must be aroused by the discovery of a buried body uninfested with maggots and which is claimed to have been exposed above ground for a number of days prior to burial in the summer. In the words of the great French entomologist, Jean Henri Fabre:

"At the surface of the soil, exposed to the air, the hideous invasion is possible; ay, it is the invariable rule. For the melting down and remoulding of matter, man is no better, corpse for corpse, than the lowest of the brutes. Then the Fly exercises her rights and deals with us as she does with any ordinary animal refuse. Nature treats us with magnificent indifference in her great regenerating-factory: placed in her crucibles, animals and men, beggars and kings are one and all alike. There you have true equality, the only equality in this world of ours: equality in the presence of the maggot".

Although exposed corpses will inevitably attract blowflies at the right time of the year, it is usually accepted that burial, even under only 2 or 3 cm of soil, will effectively prevent blowflies from laying. The very few experiments to test this statement have been done on either *Calliphora vicina* or *Calliphora vomitoria*. Some flies, other than blowflies, are known to lay their eggs on the soil above a buried body, the hatching larvae burrowing into the soil and feeding on the body. No blowfly is known to do this, but it is conceivable that some might be able to do so. Experiments with different species, using mouse carcases buried in different kinds of soil and at different depths, may well reveal some interesting facts. Is there a difference in the fly's behaviour if the soil is loosely- or hard-packed? Is there a difference in response to dry or wet soil?

Finally, the distribution of the maggots on the body of a murder victim sometimes gives a clue to the cause of death. As we have seen, blowflies normally lay their eggs in the natural body openings (such as the nose or eyes). Any departure from this pattern, such as the presence of maggots in the chest and their absence from the body openings, would suggest that a wound had been inflicted on the chest, the resulting bleeding having attracted the flies to lay their eggs in that part of the body. The actual wound may no longer be visible, but the position of the maggots would suggest that a wound was present, and therefore imply a murderous assault rather than a natural death.

Much background information on forensic entomology may be obtained from Nuorteva (1977) and Smith (1986). Further suggestions for research in forensic entomology are given by Erzinçlioğlu (1986b).

3.6 Archaeology

Unlike beetles and certain other insects, adult flies are not very robust. Consequently, they are not usually found preserved in archaeological deposits. On the other hand, the puparia of flies are very durable objects and are often found during archaeological excavations in many parts of the world. They have even been found in the wrappings of Egyptian mummies. In Britain, puparia have been found in large numbers in the Viking deposits in York, as well as in other localities such as St Kilda and Hadrian's Wall. Identification of the species of these puparia often sheds light on the environmental conditions prevailing during historic times. Such identifications could also tell us much about the relationship with man and the geographical range of blowfly species during early times, which is especially interesting if their behaviour and distribution have changed since then. Some archaeological sites, while yielding literally millions of other puparia, contain few or no blowfly puparia. It has been suggested that the human inhabitants of such localities may have lived under austere conditions where there was little meat left over in which blowflies could breed. This seems to be borne out by other evidence. Thus, the absence of blowfly puparia may itself be a useful indication of environmental conditions. Despite the absence of blowfly puparia, adult blowflies may have been abundant at such sites, exploiting the plentiful dung for protein meals.

There are great opportunities in this field for original research, which should be conducted with archaeologists as a coordinated investigation of a particular site. Unconnected fragments of information on archaeological sites are of little value and are frequently lost.

Information on fly puparia from British archaeological sites is given by Erzinçlioğlu & Phipps (1983) and Phipps (1983, 1984).

3.7 Blowflies in history

Many scientists and scholars, such as Pliny the Younger, have been aware of the facts of fly development and metamorphosis since early historic times. In Homer's *Iliad*, written about 1000 BC, we read (in translation): "I much fear, lest with blows of flies, his brass-inflicted wounds are fil'd". Shakespeare makes several references to the behaviour of blowflies. For example, in *The Tempest* (Act 3, Scene 1) he says: "to suffer the flesh-fly blow my mouth" and again in *Loves Labor Lost* (Act 5, Scene 2): "These summer flies have blowne me full of maggot ostentation". In view of this general awareness, it is surprising to find that many learned people continued to believe that maggots arose spontaneously from the decomposing carcases in which they were found and to be unaware that maggots were the offspring of flies. Many scholars entertained other strange notions about fly development. Thus, even as late as the seventeenth century, Athanasius Kircher was able to write, in his *Twelfth Book of the Subterranean World*, the following account of fly development:

"The dead flies should be besprinkled and soaked with honey-water, and then placed on a copper-plate exposed to the tepid heat of ashes; afterwards, very minute worms, only visible through the microscope, will appear, which little by little grow wings on the back and assume the shape of very small flies, that slowly attain perfect size".

Later that century, in 1668, Francesco Redi, an Italian physician, published a book entitled *Esperienze intorno alla Generazione degli Insetti* (Experiments on the Generation of Insects) in which he established once and for all that maggots hatch from eggs laid by flies. In a series of simple experiments designed to investigate the notion of spontaneous generation, he placed some dead snakes in glass bottles, some of which he sealed and some of which he left open. He found that only those snake carcases in the open bottles became infested with maggots, and he concluded:

"...I began to believe that all worms found in the meat were derived directly from the droppings of flies, and not from the putrefaction of the meat, and I was still more confirmed in this belief by having observed that, before the meat grew wormy, flies had hovered over it, of the same kind that later bred in it".

Thus was laid to rest the notion of the spontaneous generation of flies. In the scientific world of the time – the age of Isaac Newton and the birth of the Royal Society – Francesco Redi's book had such influence that no one since has seriously disputed the matter of the origin of flies.

Blowflies must have affected human beings throughout history through their carcase-breeding and parasitic behaviour, and there is evidence that the habit of myiasis in some species may have arisen as a direct consequence of the rise of human civilisation (Erzinçlioğlu, 1989b). Be that as it may, it would be interesting to document the effect of myiasis on early societies. As far as we know, the earliest undoubted historical figure to die of myiasis was Herod Agrippa, King of Judaea, in AD 48; the story of his death is recorded both by the contemporary Jewish historian Flavius Josephus, and by Saint Luke in the Acts of the Apostles in the New Testament.

Although the abundance of flies that we believe existed during early historic times must have carried disease, it is important not to forget the reverse side of this coin. The maggots of carcase- and dung-breeding flies must consume tonnes of putrefying material each year; if it were not constantly removed from the environment the disease problem would become vastly greater. Since it is these animal remains, and not the flies themselves, that are the ultimate sources of infections, a reduction of fly numbers without a corresponding reduction in the amount of carcase and dung material would only exacerbate the situation. Therefore, early human communities, with their attendant domestic animals and the associated dung and other animal remains, may have benefited in certain ways from the abundance of flies. The run-offs from latrines discovered in the Viking excavations at Coppergate in York were found to be full of the puparia of the housefly, *Musca domestica,* testifying to the amount of faeces that must have been removed by the maggots.

The documentation of the attitudes of people in early societies to flies and their effects in everyday life would be a real contribution to the understanding of human history.

4 Identification

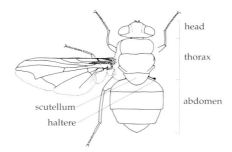

Fig. 12. A blowfly.

4.1 Adult morphology

A glance at fig. 12 will show that a blowfly, like any other insect, is divided into three parts: the head, the thorax and the abdomen. Most of the fly's head is taken up by the large compound eyes; in the males these eyes are particularly large and almost meet at the top of the head, but in the female they are separated (compare figs. 13a and b). At the front of the head is a pair of antennae. Each antenna is made up of three main segments, of which the third, or last, is by far the largest. On this last segment is a long, often feathery, hair-like structure, the arista (fig. 13). The proboscis of the fly is below the head.

The thorax bears the three pairs of legs and the wings, and houses the powerful muscles that control them. Unlike most insects, but like all true flies, blowflies have only one pair of functional wings; the hind pair is modified into club-like structures called halteres (fig. 12).

The abdomen is divided into four visible segments; it bears the genital apparatus, visible externally in the male, and contains the reproductive system. Blowflies bear an impressive array of hairs and bristles of various sizes.

Fig. 13. The head (seen from in front) of (a) a male and (b) a female blowfly, *Calliphora vomitoria*.

4.2 Using the keys

The following keys have been constructed to enable identification to be carried out easily and reliably. When identifiying a specimen do not try to rush. Read both alternatives in every couplet before looking at the specimen; the alternatives are always opposite conditions of a particular feature of the fly. Consult the figures referred to at every point and make sure you are looking at the right part of the fly.

Although the commoner bluebottles and greenbottles are familiar to most people, other blowfly species are easily confused with other flies. The insects illustrated in pl. 4 are not blowflies, although they superficially resemble them. They are common flies and it is useful to be aware of their existence in order not to confuse them with true blowflies.

Orthellia (now known as *Neomyia*) is another genus that includes greenbottle-like flies. To be sure that a fly is a calliphorid blowfly, check that it shows all the following characters.

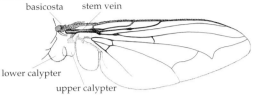

Fig. 14. The wing of a blowfly.

Fig. 15. Left lower calypter (seen from above) of a blowfly.

Fig. 16. Left lower calypter (seen from above) of a fly that is not a blowfly.

1. Lower calypter present (fig. 14), and shaped as in fig. 15, not as in fig. 16.
2. Hypopleural bristles present (fig. 17).
3. Post-scutellum not well-developed (fig. 18).
4. Post-humeral bristle at a lower level than pre-sutural (fig. 17).

Van Emden (1954) and Rognes (1991) give further details on the identification of blowflies.

4.3 The keys

Most blowflies you are likely to come across will be associated with carrion. If the flies you collected were from a carcase start with Key I, which deals only with species that frequent carrion. If flies were collected with a net or in an unbaited trap, start with Key II, which includes all known British genera.

For the species of *Lucilia* and *Bellardia*, only the males are keyed below. The females are omitted because they are very difficult to identify with certainty. When identifying specimens of these two genera, therefore, make sure the specimen you have is a male (fig. 13).

If you come across a small carcase with maggots in it, the carcase can be collected and put in a jar on a layer of sawdust or sand. (Take care – use protective rubber gloves when handling dead animals.) Place the jar in a cage and allow the maggots to develop, pupate and emerge as adult

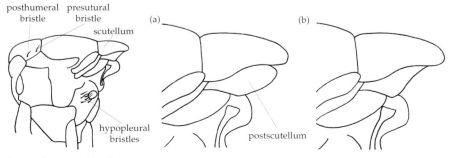

Fig. 17. Thorax of a blowfly, seen from the left.

Fig. 18. Detail of the rear of a thorax, seen from the left as in fig. 17, showing that the postscutellum may be (a) present or (b) absent.

flies. The flies can then be identified. Alternatively, it is possible to identify the maggots themselves. This is a more difficult task than adult identification, but a simple key is given below (Key VII). The following articles should be of use in further larval identification: Erzinçlioğlu (1985, 1987*a*, 1987*b*, 1988) and Zumpt (1965). For egg identification see Erzinçlioğlu (1989). Many of the characters of the immature stages require examination with the electron microscope, so a full treatment cannot be given here.

I.1

Most of the characters used in the following keys can be seen with an ordinary hand-lens, although a stereoscopic binocular microscope is preferable. However, examination of the genitalia requires dissection, and in some cases slide-mounting and examination under a compound microscope (technique, p. 55).

I Genera of blowflies frequenting carrion

1   Stem vein hairy on upper surface (fig. 14, I.1)
       Phormia (go to key VI)

–   Stem vein bare on upper surface   2

2   Suprasquamal ridge with tuft of hairs (I.2)
       Lucilia (go to key IV)

–   Suprasquamal ridge with no such tuft of hairs   3

3   Parafrontal areas (fig. 13) brilliant golden (pl. 3.8)
       Cynomya (1 species: *C. mortuorum*)

–   Parafrontal areas darker, orange or nearly black (pls. 3.1–3.3)   *Calliphora* (go to key III)

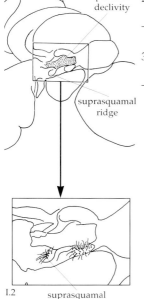

postalar declivity

suprasquamal ridge

I.2   suprasquamal ridge

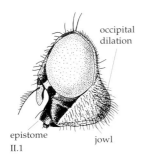

occipital
dilation

epistome   jowl

II.1

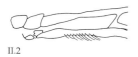

II.2

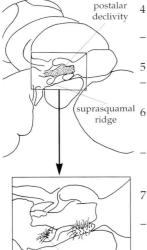

postalar
declivity

suprasquamal
ridge

suprasquamal
ridge

II.3

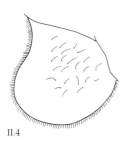

II.4

II All British blowfly genera

1 Crinkly golden hairs on upper surface and sides of
   thorax (sometimes these hairs are rubbed off on the
   upper surface which gives it a very black appearance;
   however, the hairs on the periphery and the sides are
   usually obvious) (pl. 2.2)   *Pollenia*
                     (7 species not keyed here; see chapter 2)
– No such hairs on upper surface or sides of thorax   2

2 Epistome elongated (fig. II.1)   *Stomorhina*
                                    (1 species: *S. lunata*)
– Epistome not elongate, shaped as pls. 3.1–3.8   3

3 Stem vein (fig. 14) bare on upper surface   4
– Stem vein hairy on upper surface (II.2)   10

4 Suprasquamal ridge with tuft of hairs (II.3)   *Lucilia*
                             (7 species) (Key IV, for males)
– Suprasquamal ridge with no such tuft of hairs   5

5 Lower calypter (fig. 14) hairy on upper surface (II.4) 6
– Lower calypter bare on upper surface   8

6 Parafacial area (fig. 13) with long hairs over most of its
   length (II.5)   *Bellardia*
                             (4 species) (Key V, for males)
– Parafacial area with fine hairs, if present, restricted to
   upper half (II.6)   7

7 Parafrontal areas (fig. 13) brilliant golden (pl. 3.8)
                     Cynomya (1 species: *C. mortuorum*)
– Parafrontal areas darker, orange or nearly black
   (pls. 3.1–3.3)   *Calliphora*
                             (6 species; go to key III)

8 Upper surface of abdomen brownish or blackish   9
– Upper surface of abdomen metallic blue   *Melinda*
                                   (1 species: *M. gentilis*)

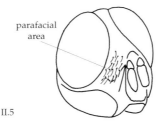

parafacial
area

II.5

II.6

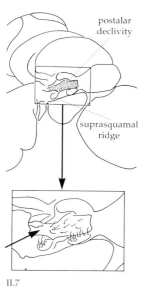

II.7

9 Postalar declivity hairy (II.7)
Pseudonesia
(1 species: *P. puberula*)

– Postalar declivity bare
Eggisops
(1 species: *E. pecchiolii*)

10 Lower calypter broad (II.8)
Phormia
(2 species) (Key VI)

– Lower calypter narrow (II.9)
Protocalliphora
(1 species: *P. azurea*)

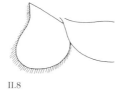

II.8

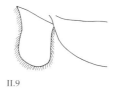

II.9

III Species of *Calliphora*

1 Lower calypter mostly dark, but pale at the margins   2
– Lower calypter wholly white or creamy   5

2 Basicosta (fig. 14) yellow   *C. vicina*
– Basicosta black   3

3 Hairs on jowls (fig. 13) yellow or orange (pl. 3.2)
C. vomitoria

– Hairs on jowls black (pl. 3.3)   4

4 Occipital dilation (see II.1) blackish (pl. 3.6)   *C. loewi*
– Occipital dilation reddish (pl. 3.3)   *C. uralensis*

5 Scutellum (fig. 12) with four or five marginal bristles each side (III.1)   *C. subalpina*
– Scutellum with only three marginal bristles each side (III.2)   *C. alpina*

III.1

III.2

IV.1

IV Males of the species of *Lucilia*

1   Basicosta (fig. 14) yellowish or creamy white   2
–   Basicosta dark orange, brown or black   3

IV.2

2   The space between the eyes (seen from above),
    more than twice width of third antennal segment (IV.1)
                                                    L. sericata

–   Eyes narrowly separated, by not more than
    width of third antennal segment (IV.2)   *L. richardsi*

3   Second abdominal segment, seen from above,
    with strong marginal bristles (IV.3)   4
–   Second abdominal segment, seen from above,
    without marginal bristles   5

IV.3

4   Surstyli tapering (IV.4)   *L. silvarum*
–   Surstyli blunt (IV.5)   *L. bufonivora*

5   Surstyli blunt (IV.6)   *L. ampullacea*
–   Surstyli tapering (IV.7 and IV.8)   6

6   Surstyli straight and forked (IV.7)   *L. caesar*
–   Surstyli curved and not forked (IV.8)   *L. illustris*

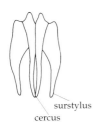

surstylus
cercus

IV.4 Male genitalia (see p. 55)

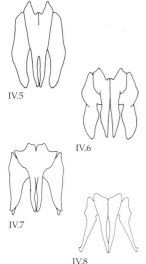

IV.5

IV.6

IV.7

IV.8

V.1

V.2

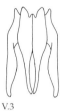

V.3

V.4

V Males of the species of *Bellardia*

1   Cerci and surstyli broad and short (V.1)   *B. cognata*
–   Cerci and surstyli narrower and longer (V.2–V.4)   2

2   Surstyli curve outwards (V.2)   *B. pusilla*
–   Surstyli curve inwards   3

3   Cerci and surstyli more elongate, and cerci with
    prominent horns (V.3)   *B. agilis*
–   Cerci and surstyli shorter, and cerci without
    prominent horns (V.4)   *B. biseta*

VI Species of *Phormia*

1   Upper calypter (fig. 14) dark, with dark hairs
                                        P. terraenovae
                                        (head, pl. 3.7)
–   Upper calypter white, with white hairs   *P. regina*
    (introduced from North America)

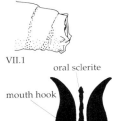

VII.1

oral sclerite

mouth hook

VII.2

VII.3

VII.4

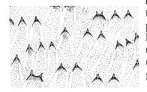

VII.5

VII.6

VII Partial key to the third-instar larvae of British carrion-breeding species

1   Posterior papillae long (seen on puparium in pl. 3.10)
                                                                Phormia

–   Posterior papillae short (VII.1)   2

2   Oral sclerite wholly black (VII.2)
                                          Calliphora or *Cynomya* 4

     Cynomya mortuorum is a fly of the north and of upland
     regions and is nowhere very common in Britain. In the
     south, it sometimes occurs in coastal areas. The great
     majority of specimens keying out here will be *Calliphora*.

–   Oral sclerite not wholly black   3

3   Oral sclerite wholly colourless (may be invisible)
     (VII.3)   *Lucilia* species
                                              (other than *ampullacea*)

–   Oral sclerite colourless, but with hindmost tip black or
     brown   *Lucilia ampullacea*

4   Mouth-hook smoothly curved (VII.2)   *Calliphora* 5

–   Mouth-hook straight, but abruptly
     curved at tip (VII.4)   *Cynomya mortuorum*

5   Body spines large, with rounded tips (VII.5)
                                                  Calliphora vomitoria

–   Body spines smaller, with pointed tips (VII.6)
                                  Calliphora vicina and other *Calliphora* species

In the south of England the species keying out here will
almost always be *Calliphora vicina* and it will also be common
in most parts of the north; however, the remaining four
Calliphora species also occur in the north and it is not possible
to separate them easily from *C. vicina*.

4.4 Synonyms

It is important to be aware that insects are often referred to
by more than one scientific name. Although there is more
than one synonym for most British blowfly species, in
practice there are only two species whose synonyms are
likely to be met with. These are: *Calliphora erythrocephala*
(now known as *Calliphora vicina*) and *Calliphora stelviana*
(called *Calliphora alpina* here). A checklist of British species is
given on p. 59.

5 Techniques

5.1 Collecting blowflies

Collecting carrion blowflies is a straightforward matter. All that is required is some bait, such as a dead mouse or a piece of meat or liver, placed in or under some kind of trap. Many complicated designs for traps have been published, but it is easy to make an effective one from ordinary household materials. The simplest trap, known as the Irwin trap, can be made from an empty plastic soft drink bottle. The top end is cut off with scissors. It is then turned round and inserted back into the main part of the bottle to form a funnel-shaped entrance. Bait may be placed at the far end and the funnel secured in position with heavy-duty sticky tape. The trap may then be placed on its side in the desired location. Most flies entering will be trapped, although some may escape. A more elaborate trap can be made from a plastic flower pot. Two strips of plastic, about 1 cm wide, are cut out near the upper rim of the pot. A hole, about 3 cm in diameter, is then made in the base of the pot. An Irwin bottle is then stuck on over the hole using heavy-duty sticky tape. The finished trap looks like fig. 19. The flower pot is then placed upside down over the bait; it is advisable to secure it in position by tying it down with string and pegs as one would do with a tent, or simply with stones. Flies will enter the trap through the slits near what is now the bottom of the trap, and will fly upwards towards the light and become trapped in the Irwin bottle. This kind of trap will catch flies more efficiently than an Irwin trap on its own. Other traps may easily be improvised.

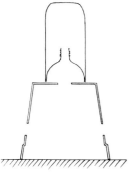

Fig. 19. A flower-pot trap.

To remove the flies from the trap, the Irwin bottle is removed and cotton wool inserted in the funnel. If the flies are required alive, the whole bottle may be placed in an insect cage (described below); the bottle can then be dismantled and the flies released inside the cage. Individual flies can be caught using small tubes. If the flies are required dead, the bottle may be placed in a refrigerator (not a freezer) for about ten minutes. The flies may then be removed into a killing jar with ethyl acetate as a killing agent.

A killing jar may be made by pouring some plaster of Paris into a clean jam jar and allowing it to set and to dry. When it is to be used, a few drops of ethyl acetate are added, and allowed to be absorbed into the plaster of Paris. It is important not to have any free liquid on the sides of the jar, or the flies may be wetted and stick to the glass; the vapour emanating from the plaster is enough to kill the flies.

Flies may be collected from naturally-occurring carcases with a butterfly net. Blowflies are strong fliers, and some practice will be required before they can be caught easily.

Non-carrion-breeding blowflies are not easy to trap. Parasitic species are best collected by finding the known host and rearing the adults. For example, *Protocalliphora azurea*

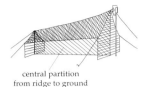

central partition
from ridge to ground

Fig. 20. A malaise trap.

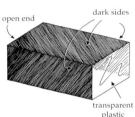

open end   dark sides

transparent
plastic

Fig. 21. A Herting trap is a
'corridor' of dark material
with one end open and the
other closed with transparent
plastic.

may be collected by removing birds' nests (immediately after
the young have flown) into a cage and awaiting the
emergence of the adults, or snail-parasites may be acquired
by collecting and keeping the snail hosts. To collect species
whose biology is unknown it is necessary to visit localities
where they are known to occur and collect them with a net
when seen, or to trap them in an unbaited trap. The Malaise
trap is widely used by entomologists. It is essentially a tent-
like structure with a flap of material suspended down the
middle. One end of the tent is higher than the other, and at
that end a bottle is attached. The flying insects land on the
flap and crawl up towards the high part of the tent,
eventually entering the collecting bottle (fig. 20). The bottle
may contain 70% alcohol (or 70% industrial methylated
spirit), or, if the insects are to be kept alive, a one-way valve.
A detailed description of a Malaise trap is given by Cogan &
Smith (1984).

The Herting trap is a useful trap for catching strong-
flying insects like blowflies alive. This is, in effect, a corridor
of dark material. One end is open and the other is closed
with a window of transparent plastic sheeting (fig. 21). The
flies fly into the corridor towards the light, eventually
landing upon the plastic sheeting. They will usually stay
there, since they are reluctant to fly back into the dark
corridor. The flies can be collected in specimen tubes from
the window.

5.2 Rearing methods

It is very easy to rear carrion blowflies in cages. A
simple cage can be made by bending thick wire to form a
cubical framework with sides about 30 cm long, and
covering this with a length of tubular knitted cotton
dishcloth material, knotted closed at each end.

To start a colony, gravid females, or males and females,
are introduced into the cage. A piece of liver in a small dish
is placed in the cage; if the females are ready they will soon
lay eggs on the liver. If they are not, they will feed on it to
mature their eggs. About a week or ten days after their first
protein meal, the flies will be ready to lay eggs. When a
suitable egg mass appears on the liver, the dish is moved into
a plastic box containing sawdust or clean sand; more liver is
then added to the dish. The liver must not be allowed to dry
out. The larvae will hatch and feed on the liver; when they
have finished feeding they will leave the dish and pupate in
the sawdust or sand. Most kinds of mammal or bird meat are
suitable for rearing blowflies, but heart-meat is not. It lacks
vitamin A, which is needed by the flies to synthesise the
rhodopsin necessary for vision. Flies emerging from larvae
reared on heart-meat will not behave normally.

When the larvae have pupated, the dish with the
remains of the meat is removed and the plastic box
containing the pupae is placed in a clean cage for emergence.
Flies must be kept supplied with clean water and sugar at all
times, by providing water-soaked cotton wool in a small

bowl, and dry sugar lumps in a dish. As soon as the flies emerge in the new cage, a new piece of liver for protein feeding is introduced as above.

Parasitic blowfly species are not easy to rear in captivity, and there is scope for pioneering work in developing such techniques.

5.3 Preservation

Adults or larvae to be dissected are best preserved in acetic alcohol (3 parts 80% ethyl alcohol: 1 part glacial acetic acid). A more readily available alternative is 70% industrial methylated spirit with a little glycerol to slow evaporation. Preservation in alcohol alone is not recommended, because the specimens become brittle and difficult to dissect; acetic acid keeps the tissues soft.

Adult blowflies for a reference collection are pinned and stored dry. It is advisable to pin the flies as soon as possible after killing them, when they are still flexible enough for their wings and legs to be arranged in the desired positions. At this stage it is often useful to expose the male genitalia by pulling them out from the tip of the abdomen with a tiny hooked pin, in case they are needed for identification. The best way to mount blowfly specimens is to stage them. Using a fine, headless, stainless steel entomological pin, the fly is mounted on a small a strip of polyethylene foam (a white synthetic substitute for cork). The small pin carrying the blowfly itself should not have anything else attached to it. The strip of foam itself may then be mounted on using an ordinary (e.g. no. 12) entomological pin. Labels bearing information about the time and place of capture, as well as the name of the species, are pinned lower down on the same large pin. The large pin, bearing the foam strip with its staged fly, is stuck into a sheet of foam lining the base of an insect box or a drawer in a cabinet.

5.4 Examining flies

A binocular stereoscopic microscope is desirable for the identification of blowflies, but a strong lens will suffice in many cases. The identification of the males of *Lucilia* species will require examination of the genitalia. It is rarely possible to see enough detail of these without dissection. This may be done by placing the fresh or wet-preserved specimen on a microscope slide and cutting off the hind third of the abdomen with a sharp scalpel. The genital segments can then be dissected out in a drop of fluid using pins, keeping the material wet all the time. This operation is not as difficult as it may sound, since the parts used for identification (see keys) are very distinctive and easy to find. The dissected parts can then be mounted on a clean slide in a drop of water under a coverslip, and examined with a compound microscope. If a more permanent preparation is required, the genitalia may be mounted in a few drops of Berlese's Fluid under a coverslip. The slide may be placed on a hot-plate or

near a radiator to dry, and it may be necessary to add further
Berlese's Fluid around the edges of the coverslip to make up
for evaporation losses. Berlese's Fluid has the advantage that
permanent mounts can be made very quickly, and that, being
water-based, it can be dissolved away if necessary so that the
specimen can be re-mounted if so desired. A number of
formulae for Berlese's Fluid exist in the literature; some of
these give incorrect proportions of the various components.
The following is the formula that should be used.

Gum arabic, picked lumps	12g
Chloral hydrate crystals	20g
Glacial acetic acid	5ml
50% w/w glucose syrup	5ml
Distilled water	30–40ml

The components should be dissolved in the distilled
water in the order shown. The mixture is then filtered
through glass wool to remove dust.

5.5 Marking flies

Flies trapped as described in section 5.1 are released
into a cage, from which they can be removed individually,
using small glass specimen tubes. They are marked by
applying a spot of paint (such as Humbrol model aeroplane
paint, hobby paint or artists' oil paint) to the abdomen with a
fine brush. This can be difficult, but, with practice, it should
be possible to insert the brush into the tube and daub the
paint onto the fly.

5.6 Statistical analysis

Sometimes it may be necessary to subject the results of
your investigation to statistical tests. In a book this size it is
not possible to describe all the statistical tests that you may
need to use, and the project guide by Chalmers & Parker
(1989) is recommended for this purpose. The book by
Southwood (1978) is a useful source-book on ecological
methods generally.

5.7 Publishing your findings

There is little use in carrying out original and
interesting work if the results are not published. Articles or
notes for publication in journals should be written concisely
in a clear, straightforward style. Short notes reporting, say,
the discovery of a species in a new geographical location or
some new natural history observations, can be written as a
simple essay. Such notes can be sent to such journals as *The
Entomologists' Monthly Magazine, The Entomologists' Record
and Journal of Variation, The Entomologist* or *The Naturalist.*
Longer, more detailed reports of investigations should be
divided into Introduction, Materials and Methods, Results
and Discussion sections. It is sometimes more convenient to

include the Results and Discussion together as one section. Conventionally, the general background and justification of the work are explained in the Introduction. The equipment and techniques used are described in the Materials and Methods section in sufficient detail to allow anyone wishing to repeat the work to do so easily. It is important to refer to published articles relating to the work you have carried out; it is necessary to cite such publications when your work supports or contradicts previous findings.

Further reading

Introductory books, such as Colyer and Hammond's *Flies of the British Isles,* may be readily available from your local public library. More specialised books, or articles in scientific journals, may have to be ordered by the local library from the British Library, Document Supply Centre, either as loans or as photocopies. If you live in a university town it may be possible to register as a reader at the university library.

Anderson, D.S. (1960). The respiratory system of the egg-shell of *Calliphora erythrocephala. Journal of Insect Physiology* **5**, 120–128.

Askew, R.R. (1978). *Parasitic Insects.* London: Heinemann.

Barton Browne, L. (1958). The choice of communal oviposition sites by the Australian sheep blowfly, *Lucilia cuprina. Australian Journal of Zoology* **6**, 241–247.

Barton Browne, L. (1962). The relationship between oviposition in the blowfly *Lucilia cuprina* and the presence of water. *Journal of Insect Physiology* **8**, 383–390.

Barton Browne, L., Bartell, R.J. & Shorey, H.H. (1969). Pheromone-mediated behaviour leading to group oviposition in the blowfly, *Lucilia cuprina. Journal of Insect Physiology* **15**, 1003–1014.

Barton Browne, L., van Gerwen, A.C.M. & Bartell, R.J. (1980). Effects of the ingestion of components of a protein rich diet on the sexual receptivity of females of the Australian sheep blowfly, *Lucilia cuprina. Physiological Entomology* **5**, 1–6.

Beveridge, W.I.B. (1935). Urine soiling in ewes in relation to blowfly strike. *Australian Veterinary Journal* **11**, 104–106.

Blackith, R.E. & Blackith, R.M. (1984). Larval aggression in Irish flesh-flies. *Irish Naturalists' Journal* **21**, 255–257.

Blackith, R.E. & Blackith, R.M. (1990). Insect infestations of small corpses. *Journal of Natural History* **24**, 699–709.

Block, W., Erzinçlioğlu, Y.Z. & Worland, R.P. (1990). Cold resistance in all life stages of two blowfly species (Diptera: Calliphoridae). *Medical and Veterinary Entomology* **4**, 213–219.

Bolwig, N. (1946). Senses and sense organs of the anterior end of the house fly larvae. *Videnskabelige Meddelelser fra Dansk naturhistorik Forening Kjøbenhavn* **109**, 81–217.

Broce, A.B. (1980). Sexual behaviour of screw-worm flies stimulated by swormlure – 2. *Annals of the Entomological Society of America* **73**, 386–389.

Campan, M. & Langa, J. (1981). Étude expérimentale des enchainments d'actes dans le comportement sexuel de *Calliphora vomitoria*: approches éthologique et statistique. *Biology and Behaviour* **6**, 19–34.

Chalmers, N. & Parker, P. (1989). *The OU Project Guide. Fieldwork and Statistics for Ecology Projects.* (2nd edition). London: Open University / Field Studies Council.

Chernoguz, D.G. (1984). An ontogenetic aspect of parasitism of the parasitoids of the subfamily Alysiinae (Hymenoptera, Braconidae). Relation of *Alysia manducator* Panz. and *Calliphora vicina* R.-D. *Entomologicheskoe Obozrenie* **63**, 658–671.

Cogan, B.H. & Smith, K.G.V. (1984). *Insects: instructions for collectors.* London: British Museum (Natural History).

Colyer, C.N. & Hammond, C.O. (1968). *Flies of the British Isles.* London: Frederick Warne & Co. Ltd.

Coope, G.R. & Lister, A.M. (1987). Late glacial mammoth skeletons from Condover, Shropshire, England. *Nature, London* **330**, 587.

Corbet, S.A. (1985). Insect chemosensory responses: a chemical legacy hypothesis. *Ecological Entomology* **10**, 143–153.

Corbet, S.A., Sellick, R.D. & Willoughby, N.G. (1974). Notes on the biology of the mayfly *Povilla adusta* in West Africa. *Journal of Zoology* **172**, 491–502.

Cragg, J.B. (1950*a*). Studies on *Lucilia* species (Diptera) under Danish conditions. *Annals of Applied Biology* **37**, 66–79.

Cragg, J.B. (1950*b*). The reactions of *Lucilia sericata* (Mg.) to various substances placed on sheep. *Parasitology* **40**, 179–186.

Cragg, J.B. (1955). The natural history of sheep blowflies in Britain. *Annals of Applied Biology* **42**, 197–207.

Cragg, J.B. (1956). The olfactory behaviour of *Lucilia* species (Diptera) under natural conditions. *Annals of Applied Biology* **44**, 467–477.

Cragg, J.B. & Cole, P. (1956). Laboratory studies on the chemosensory reactions of blowflies. *Annals of Applied Biology* **44**, 478–491.

Cragg, J.B. & Ramage, G.R. (1945). Chemotropic studies on the blow-fly *Lucilia sericata* (Mg.) and *Lucilia caesar* (L.). *Parasitology* **36**, 168–175.

Cragg, J.B. & Thurston, B.A. (1950). The reactions of blowflies to organic sulphur compounds and other materials used in traps. *Parasitology* **40**, 187–194.

Dafni, A. (1984). Mimicry and deception in pollination. *Annual Review of Ecology and Systematics* **15**, 259–278.

Dallwitz, R. (1987). Density independence of survival in myiasis breeding of *Lucilia cuprina*. *Bulletin of Entomological Research* **77**, 171–176.

Davidson, J. (1918). Some practical methods adopted for the control of flies in the Egyptian campaign. *Bulletin of Entomological Research* **8**, 297–309.

Davies, L. (1948). Observations on the development of *Lucilia sericata* (Mg.) eggs in sheep fleeces. *Journal of Experimental Biology* **25**, 86–102.

Davies, L. (1950). The hatching mechanism of Muscid eggs. *Journal of Experimental Biology* **27**, 437–445.

Davies, L. (1990). Species composition and larval habitats of blowfly (Calliphoridae) populations in upland areas in England and Wales. *Medical and Veterinary Entomology* **4**, 61–68.

Davies, W.M. (1929). Hibernation of *Lucilia sericata*, Mg. *Nature, London* **123**, 759–760.

Davies, W.M. & Hobson, R.P. (1935). Sheep blowfly investigations. I. The relationship of humidity to blowfly attack. *Annals of Applied Biology* **22**, 279–293.

Dear, J.P. (1986). Calliphoridae (Insecta, Diptera). *Fauna of New Zealand* **8**, 1–86. Wellington: D.S.I.R.

Denno, R.F. & Cothran, W.R. (1975). Niche relationships of a guild of necrophagous flies. *Annals of the Entomological Society of America* **68**, 741–754.

Denno, R.F. & Cothran, W.R. (1976). Competitive interactions and ecological strategies of sarcophagid and calliphorid flies inhabiting rabbit carrion. *Annals of the Entomological Society of America* **69**, 109–113.

Dethier, V. (1976). *The Hungry Fly. A Physiological Study of the Behaviour Associated with Feeding.* Cambridge, Mass. & London: Harvard University Press.

Dicke, R.J. & Eastwood, J.P. (1952). The seasonal incidence of blowflies at Madison, Wisconsin. *Wisconsin Academy of Sciences, Arts and Letters Transactions* **41**, 207–217.

Digby, P.S.B. (1958*a*). Flight activity in the blowfly *Calliphora erythrocephala* in relation to light and radiant heat, with special reference to adaptation. *Journal of Experimental Biology* **35**, 1–19.

Digby, P.S.B. (1958*b*). Flight activity of the blowfly, *Calliphora erythrocephala*, in relation to wind speed, with special reference to adaptation. *Journal of Experimental Biology* **35**, 776–795.

Disney, R.H.L. (1973). Some flies associated with dog dung in an English city. *Entomologist's Monthly Magazine* **108**, 93–94 (1972).

Disney, R.H.L. (1986). Morphological and other observations on *Chonocephalus* (Phoridae) and phylogenetic implications for the Cyclorrhapha (Diptera). *Journal of Zoology* **210**, 77–87.

Disney, R.H.L., Erzinçlioğlu, Y.Z., Henshaw, D. de C., Howse, D., Unwin, D.M., Withers, P. & Woods, A. (1982). Collecting methods and the adequacy of attempted fauna surveys, with reference to the Diptera. *Field Studies* **5**, 607–621.

Dodge, H.R. (1952). A possible case of blowfly myiasis in a rat, with notes on the bionomics of *Bufolucilia silvarum*. *Entomological News* **63**, 212–214.

Edwards, D.K. (1961). Activity of two species of *Calliphora* (Diptera) during barometric pressure changes of natural magnitude. *Canadian Journal of Zoology* **39**, 623–635.

Eisemann, C.H. & Rice, M.J. (1987). The origin of sheep blowfly, *Lucilia cuprina* (Wiedemann) (Diptera: Calliphoridae), attractants in media infested with larvae. *Bulletin of Entomological Research* **77**, 287–294.

Emden, F. I. van (1954). Diptera Cyclorrhapha Calyptrata (I). Section (*a*). Tachinidae and Calliphoridae. *Handbooks for the Identification of British Insects* **10**(4*a*), 1–133. London: Royal Entomological Society of London.

Emmens, R.L. (1981). Evidence for an attractant in cuticular lipids of female *Lucilia cuprina* (Wied.), Australian sheep blowfly. *Journal of Chemical Ecology* **7**, 529–541.

Emmens, R.L. & Murray, M.D. (1982). The role of bacterial odours in oviposition by *Lucilia cuprina* (Wiedemann) (Diptera: Calliphoridae), the Australian sheep blowfly. *Bulletin of Entomological Research* **72**, 367–375.

Emmens, R.L. & Murray, M.D. (1983). Bacterial odours as oviposition stimulants for *Lucilia cuprina* (Wiedemann)(Diptera: Calliphoridae), the Australian sheep blowfly. *Bulletin of Entomological Research* **73**, 411–415.

Erzinçlioğlu, Y.Z. (1981). On the Diptera associated with dog-dung in London. *London Naturalist* **60**, 45–46.

Erzinçlioğlu, Y.Z. (1984). A new parasite record for *Protocalliphora azurea* (Fall.) (Dipt., Calliphoridae). *Entomologist's Monthly Magazine* **120**, 172.

Erzinçlioğlu, Y.Z. (1985). Immature stages of British *Calliphora* and *Cynomya*, with a re-evaluation of the taxonomic characters of larval Calliphoridae (Diptera). *Journal of Natural History* **19**, 69–96.

Erzinçlioğlu, Y.Z. (1986*a*). An experiment with carrion flies in Hayley Wood. *Nature in Cambridgeshire* **28**, 9–12.

Erzinçlioğlu, Y.Z. (1986*b*). Areas of research in forensic entomology. *Medicine, Science and the Law* **26**, 273–278.

Erzinçlioğlu, Y.Z. (1987*a*). Recognition of the early instar larvae of the genera *Calliphora* and *Lucilia* (Dipt., Calliphoridae). *Entomologists' Monthly Magazine* **123**, 97–98.

Erzinçlioğlu, Y.Z. (1987*b*). The larvae of some blowflies of medical and veterinary importance. *Medical and Veterinary Entomology* **1**, 121–125.

Erzinçlioğlu, Y.Z. (1988). The larva of the species of *Phormia* and *Boreellus*: Northern, cold-adapted blowflies (Diptera: Calliphoridae). *Journal of Natural History* **22**, 11–16.

Erzinçlioğlu, Y.Z. (1989*a*). Entomology and the forensic scientist: how insects can solve crimes. *Journal of Biological Education* **23**, 300–302.

Erzinçlioğlu, Y.Z. (1989*b*). The origin of parasitism in blowflies. *British Journal of Entomology and Natural History* **2**, 125–127.

Erzinçlioğlu, Y.Z. (1990). Entomology and the forensic scientist: how insects can solve crimes. *Journal of Biological Education* **23**, 300–302.

Erzinçlioğlu, Y.Z. & Davies, S.W. (1984). The blue-bottle fly *Calliphora vicina* R.-D. as a parasite (primary myiasis agent), particularly in small mammals. *Naturalist* **109**, 31–34.

Erzinçlioğlu, Y.Z. & Phipps, J. (1983). In: Hall, A.R., Kenward, H.K., Williams, D. & Greig, J.R.A. Environment and living conditions at two Anglo-Scandinavian sites. *The Archaeology of York* **14/4**, 157–240.

Erzinçlioğlu, Y.Z. & Whitcombe, R.P. (1983). *Chrysomya albiceps* (Wiedemann) (Dipt., Calliphoridae) in dung and causing myiasis in Oman. *Entomologist's Monthly Magazine* **119**, 51–52.

Evans, A.C. (1933). Comparative observations on the morphology and biology of some hymenopterous parasites of carrion-infesting Diptera. *Bulletin of Entomological Research* **24**, 385–405.

Evans, A.C. (1935). Some notes on the biology and physiology of the sheep blowfly, *Lucilia sericata*, Meig. *Bulletin of Entomological Research* **26**, 115–122.

Evans, A.C. (1936). Studies on the influence of the environment on the sheep blow-fly *Lucilia sericata* Meig. IV. The indirect effect of temperature and humidity acting through certain competing species of blow-flies. *Parasitology* **28**, 431–439.

Fabre, J.H. (1919). *The Life of the Fly.* London: Hodder & Stoughton.

Fraenkel, G. (1935). Observations and experiments on the blow-fly (*Calliphora erythrocephala*) during the first day after emergence. *Proceedings of the Zoological Society of London* **1935**, 893–904.

Fraenkel, G. & Bhaskarian, G. (1973). Pupariation and pupation in cyclorrhaphous flies (Diptera): terminology and interpretation. *Annals of the Entomological Society of America* **66**, 418–422.

Free, J.B. (1993). *Insect Pollination of Crops.* London: Academic Press.

French, N., Wall, R., Cripps, P.J. & Morgan, K.L. (1994). Blowfly strike in England and the relationship between prevalence and farm management factors. *Medical and Veterinary Entomology* **8**, 51–56.

Freney, M.R. (1937). Studies on the chemotropic behaviour of sheep blowflies. *Council for Scientific and Industrial Research, Australia,* Pamphlet No. **74**, 1–24.

Gauld, I. & Bolton, B. (1988). *The Hymenoptera.* London: British Museum (Natural History)/Oxford University Press.

Gautier, A. (1975). Fossiele Vliegenmaden *(Protophormia terraenovae* (Robineau-Desvoidy, 1830)) in een Schedel van de wolharige Neushoorn *(Coelodonta antiquitatis)* uit het Onder-Wurm te Dendermonde (Oost-Vlaanderen, Belgie). *Natuurwetenschappelijk Tijdschrift* **56**, 76–84.

Gautier, A. & Schumann, H. (1973). Puparia of the subarctic blowfly *Protophormia terraenovae* (Robineau-Desvoidy, 1830) in a skull of a Late Eemian (?) bison at Zemst, Brabant (Belgium). *Palaeogeography, Palaeoclimatology, Palaeoecology,* **14**, 119–125.

van Gerwen, A.C.M., Barton Browne, L., Vogt, W.G. & Williams, K.L. (1987). Capacity of autogenous and anautogenous females of the Australian sheep blowfly, *Lucilia cuprina*, to survive water and sugar deprivation following emergence. *Entomologia experimentalis et applicata* **43**, 209–214.

Graham-Smith, G.S. (1916). Observations on the habits and parasites of common flies. *Parasitology* **8**, 440–544.

Graham-Smith, G.S. (1919). Further observations on the habits and parasites of common flies. *Parasitology* **11**, 347–384.

Graham-Smith, G.S. (1930). Further observations on the anatomy and function of the proboscis of the blow fly, *Calliphora erythrocephala. Parasitology* **22**, 47–115.

Greathead, D.J. (1962). The biology of *Stomorhina lunata* (Fabricius) (Diptera: Calliphoridae) a predator of the eggs of Acrididae. *Proceedings of the Zoological Society of London* **139**, 139–180.

Green, A.A. (1951). The control of blowflies infesting slaughterhouses. 1. Field observations on the habits of blowflies. *Annals of Applied Biology* **38**, 475–494.

Greenberg, B. (1971). *Flies and Disease.* Vol. 1. N.J.: Princeton University Press.

Greenberg, B. (1973). *Flies and Disease.* Vol. 2. N.J.: Princeton University Press.

Greenberg, B. (1985). Forensic entomology: case studies. *Bulletin of the Entomological Society of America* **31**, 25–28.

Greenberg, B. (1991). Flies as forensic indicators. *Journal of Medical Entomology* **28**, 565–577.

Grinfel'd, E.K. (1955). The nectar and pollen nutrition of Diptera and their role in plant pollination. *Vestnik Leningradskogo Gosundarstvennogo Universita* **10**, 15–25.

Guerrini, V.H., Murphy, G.M. & Broadmeadow, M. (1988). The role of pH in the infestation of sheep by *Lucilia cuprina* larvae. *International Journal of Parasitology* **18**, 407–409.

Gurney, W.B. & Woodhill, A.R. (1926). Investigations on sheep blowflies. Part 1. Range of flight and longevity. *New South Wales Department of Agricultutre Scientific Bulletin* **27**, 1–19.

Gurney, W.S.C., Blythe, S.P. & Nisbet, R.M. (1980). Nicholson's blowflies revisited. *Nature, London* **287**, 17–21.

Gwatkin, R. & Fallis, A.M. (1938). Bactericidal and antigenic qualities of the washings of blow fly maggots. *Canadian Journal of Research (D) Zoology* **16**, 343–352.

Haddow, A.J. & Thomson, R.C.M. (1937). Sheep myiasis in south-west Scotland, with special reference to the species involved. *Parasitology* **29**, 96–116.

Hanski, I. (1976). Breeding experiments with carrion flies (Diptera) in natural conditions. *Annales Entomologici Fennici* **42**, 113–121.

Hanski, I. (1987). Carrion fly community dynamics: patchiness, seasonality and coexistence. *Ecological Entomology* **12**, 257–266.

Hanski, I. & Kuusela, S. (1977*a*). An experiment on competition and diversity in the carrion fly community. *Annales Entomologici Fennici* **43**, 108–115.

Hanski, I. & Kuusela, S. (1977*b*). The structure of carrion fly communities: differences in breeding seasons. *Annales Zoologici Fennici* **17**, 185–190.

Hanski, I. & Nuorteva, P. (1975). Trap survey of flies and their diel periodicity in the subarctic Kevo Nature Reserve, Northern Finland. *Annales Entomologici Fennici* **43**, 56–64.

Hedström, L. & Nuorteva, P. (1971). Zonal distribution of flies on the hill Ailigas in subarctic northern Finland. *Annales Entomologici Fennici* **37**, 121–125.

Hennig, W. (1950). Entomologische Beobachtungen an kleinen Wirbeltierleichen. *Zeitschrift für Hygienisch Zoologie* **38**, 33–88.

Hepburn, G.A. (1943). Sheep blowfly research. I. A survey of maggot collections from live sheep and a note on the trapping of blowflies. *Onderstepoort Journal of Veterinary Science and Animal Industry* **18**, 13–18.

Hepburn, G.A. & Nolte, M.C.A. (1943). Sheep blowfly research III. Studies on the olfactory reactions of sheep blowflies. *Onderstepoort Journal of Veterinary Science and Animal Industry* **18**, 27–48

Herms, W.B. (1928). The effect of different quantities of food during the larval period on the sex ratio and size of *Lucilia sericata* (Meigen) and *Theobaldia incidens* Thomson. *Journal of Economic Entomology* **21**, 720–729.

Hightower, B.G. & Alley, D.A. (1963). Local distribution of released laboratory-reared screw-worm flies in relation to water sources. *Journal of Economic Entomology* **56**, 798–802.

Hinton, H.E. (1948). On the origin and function of the pupal stage. *Transactions of the Royal Entomological Society of London* **99**, 395–409.

Hinton, H.E. (1981). *The Biology of Insect Eggs*. Oxford: Pergamon Press.

Hobson, R.P. (1932). Studies on the nutrition of blow-fly larvae. III. The liquefaction of muscle. *Journal of Experimental Biology* **9**, 359–365.

Hobson, R.P. (1935). Sheep blow-fly investigations. II. Substances which induce *Lucilia sericata* Mg. to oviposit on sheep. *Annals of Applied Biology* **22**, 294–300.

Hobson, R.P. (1936). Sheep blow-fly investigations. IV. On the chemistry of the fleece, with reference to the susceptibility of sheep to blow-fly attack. *Annals of Applied Biology* **23**, 852–861.

Holdaway, F.G. (1930). Field populations and natural control of *Lucilia sericata*. *Nature* **126**, 648–649.

Holdaway, F.G. & Mulhearn, C.R. (1934). Field observations on weather stain and blowfly strike of sheep, with special reference to body strike. *Pamphlet of the Council for Scientific and Industrial Research* **48**, 1–35.

Ibrahim, I.A. & Gad, A.M. (1978). The occurrence of paedogenesis in *Eristalis* larvae (Diptera, Syrphidae). *Journal of Medical Entomology* **12**, 286

Ibrahim, S.H. (1984). A study on dipterous parasites of honeybees. *Zeitschrift für angewandte Entomologie* **97**, 124–126.

Jones, T.H. & Turner, B.D. (1987). The effect of host spatial distribution on patterns of parasitism by *Nasonia vitripennis*. *Entomologia Experimentalis et Applicata* **44**, 169–175.

Kaib, M. (1974). Die Fleisch- und Blumen-duftrezeptoren auf der Antenne der Schmeissfliege *Calliphora vicina*. *Journal of Comparative Physiology* **95**, 105–121.

Kamal, A.S. (1958). Comparative study of thirteen species of sarcosaprophagous Calliphoridae and Sarcophagidae (Diptera). 1. Bionomics. *Annals of the Entomological Society of America* **51**, 261–271.

Keilin, D. (1919). On the life-history and anatomy of *Melinda cognata* Meigen (Diptera Calliphoridae) parasitic in the snail *Helicella (Heliomanes) virgata* da Costa, with an account of the other Diptera living upon molluscs. *Parasitology* **11**, 430–455.

Keilin, D. (1924). The absence of paedogenetic multiplication in the blow-fly (*Calliphora erythrocephala*). *Parasitology* **16**, 239–247.

Keilin, D. (1944). Respiratory systems and respiratory adaptations in larvae and pupae of Diptera. *Parasitology* **36**, 1–66.

Kirk, W.D.J. (1984). Ecologically selective water traps. *Ecological Entomology* **9**, 35–41.

Kitching, J.W. (1980). On some fossil Arthropoda from the Limeworks, Makapansgat, Potgietersus. *Palaeontologia Africana* **23**, 63–68.

Kloet, G.S. & Hincks, W.D. (1976). A Check List of British Insects (2nd edition). Part 5: Diptera and Siphonaptera. *Handbooks for the Identification of British Insects* **11**(5), 1–139. London: Royal Entomological Society of London.

Kuusela, S. & Hanski, I. (1982). The structure of carrion fly communities: the size and type of carrion. *Holarctic Ecology* **5**, 337–348.

Laing, J. (1935). On the ptilinum of the blow-fly (*Calliphora erythrocephala*). *Quarterly Journal of Microscopical Science* **77**, 497–521.

Lane, R.P. (1975). An investigation into blowfly (Diptera: Calliphoridae) succession on corpses. *Journal of Natural History* **9**, 581–588.

Lawson, J.R. & Gemmell, M.A. (1985). The potential role of blowflies in the transmission of taeniid tapeworm eggs. *Parasitology* **91**, 129–143.

Mackerras, I.M. & Mackerras, M.J. (1944). Sheep blowfly investigations. The attractiveness of sheep for *Lucilia cuprina*. *Bulletin of the Council for Scientific and Industrial Research* **181**, 1–44.

Mackerras, M.J. (1933). Observations on the life-histories, nutritional requirements and fecundity of blowflies. *Bulletin of Entomological Research* **24**, 353–362.

MacLeod, J. (1947). The climatology of blowfly myiasis. I. Weather and oviposition. *Bulletin of Entomological Research* **38**, 285–303.

MacLeod, J. (1949). The climatology of blowfly myiasis. II. Oviposition and daily weather indices. *Bulletin of Entomological Research* **40**, 179–201.

MacLeod, J. & Donnelly, J. (1956). Methods for the study of blowfly populations. II. The use of laboratory-bred material. *Annals of Applied Biology* **44**, 643–648.

MacLeod, J. & Donnelly, J. (1957). Some ecological relationships of natural populations of calliphorine blowflies. *Journal of Animal Ecology* **26**, 135–170.

MacLeod, J. & Donnelly, J. (1958). Local distribution and dispersal paths of blowflies in hill country. *Journal of Animal Ecology* **27**, 349–374.

MacLeod, J. & Donnelly, J (1960). Natural features and blowfly movement. *Journal of Animal Ecology* **29**, 85–93.

MacLeod, J. & Donnelly, J. (1962). Microgeographic aggregations in blow-fly populations. *Journal of Animal Ecology* **31**, 525–543.

MacLeod, J. & Donnelly, J. (1963). Dispersal and interspersal of blowfly populations. *Journal of Animal Ecology* **31**, 1–32.

Maxwell-Lefroy, H. (1916). The control of flies and vermin in Mesopotamia. *Agricultural Journal of India* **11**, 323–331.

McAlpine, J.F. (1970). First record of acalypterate flies in the Mesozoic era (Diptera: Calliphoridae). *Canadian Entomologist* **102**, 342–346.

Medawar, P.B. (1979). *Advice to a Young Scientist*. New York: Harper & Row.

Mellanby, K. (1938). Diapause and metamorphosis of the blowfly *Lucilia sericata* Meig. *Parasitology* **30**, 392–399.

Meskin, I. (1986). Factors affecting the coexistence of blowflies (Diptera: Calliphoridae) on the Transvaal Highveld, South Africa. *South African Journal of Science* **82**, 244–250.

Molyneux, A.S. & Bedding, R.A. (1984). Influence of soil texture and moisture on the infectivity of *Heterorhabditis* sp. and *Steinemema glaseri* for larvae of the sheep blowfly, *Lucilia cuprina*. *Nematologica* **30**, 358–365.

Munro Fox, H. & Pugh Smith, G. (1933). Growth stimulation of blow-fly larvae fed on fatigued frog muscle. *Journal of Experimental Biology* **10**, 196–200.

Murray, M.D. & Wilkinson, F.C. (1980). Blowfly strike of sheep in southern Australia. 2. Western Australia. *Agricultural Record* **7**, 50–53.

Myers, J.G. (1929). Further notes on the habits of *Alysia manducator* and other parasites (Hym.) of muscoid flies. *Bulletin of Entomological Research* **19**, 357–360.

Nicholson, A.J. (1954). Compensatory reactions of populations to stresses, and their evolutionary significance. *Australian Journal of Zoology* **2**, 1–65.

Nicholson, A.J. (1957). The self-adjustment of populations to change. *Cold Spring Harbor Symposia in Quantitative Biology* **22**, 153–173.

Nielsen, B.O. & Nielsen, S.A. (1946). Schmeissfliegen (Calliphoridae) und vakuumverpackter Schinken. *Anzeiger für Schadlingskunde Pflanzen und Umweltschutz* **49**, 113–115.

Norris, K.R. (1959). The ecology of sheep blowflies on Australia. In: *Biogeography and ecology in Australia. Monographiae Biologicae* **8**, 514–544.

Norris, K.R. (1965). The bionomics of blowflies. *Annual Review of Entomology* **10**, 47–68.

Norris, K.R. (1966). Daily patterns of flight activity of blowflies (Calliphoridae: Diptera) in the Canberra district as indicated by trap catches. *Australian Journal of Zoology* **14**, 835–853.

Nuorteva, P. (1959). Studies on the significance of flies in the transmission of poliomyelitis. III. The composition of the blowfly fauna, and the activity of the flies in relation to the weather during the epidemic season of poliomyelitis in South Finland. *Annales Entomologici Fennici* **25**, 121–136.

Nuorteva, P. (1965*a*). The flying activity of blowflies (Dipt., Calliphoridae) in subarctic conditions. *Annales Entomologici Fennici* **31**, 242–245.

Nuorteva, P. (1965*b*). Synanthropy of blowflies. *Proceedings of the XIIth International Congress of Entomology* **12**, 786.

Nuorteva, P. (1970). Histerid beetles as predators of blowflies (Diptera, Calliphoridae) in Finland. *Annales Zoologici Fennici* **7**, 195–198.

Nuorteva, P. (1972). A three year study of the duration of development of *Cynomyia mortuorum* (L.) (Dipt., Calliphoridae) in the conditions of a subarctic fell. *Annales Entomologici Fennici* **38**, 65–74.

Nuorteva, P. (1977). Sarcosaprophagous insects as forensic indicators. In: Tedeschi, C.G., Eckert, W.G. & Tedeschi, L.G. (Eds). *Forensic Medicine: A study in trauma and environmental hazards.* Vol. II, 1072–1095. Philadelphia: W.B. Saunders Company.

Nuorteva, P. & Laurikainen, E. (1964). Synanthropy of blowflies (Dipt., Calliphoridae) on the island of Gotland, Sweden. *Annales Entomologici Fennici* **30**, 187–190.

Nuorteva, P. & Räsänen, T. (1968). The occurrence of blowflies (Dipt., Calliphoridae) in the archipelago of the lake Kallavesi, Central Finland. *Annales Zoologici Fennici* **5**, 188–193.

Nuorteva, P. & Skaren, U. (1960). Studies on the significance of flies in the transmission of poliomyelitis. V. Observations on the attraction of blowflies to the carcases of micromammals in the commune of Kuhmo, East Finland. *Annales Entomologici Fennici* **26**, 221–226.

Parker, G.A. (1968). The sexual behaviour of the blowfly, *Protophormia terraenovae* R.-D. *Behaviour* **32**, 291–308.

Patten, B.M. (1914). A quantitative determination of the orienting reaction of the blowfly larva (*Calliphora erythrocephala* Meigen). *Journal of Experimental Zoology* **17**, 213–280.

Phipps, J. (1983). Looking at puparia. *Circaea* **1**, 13–29.

Phipps, J. (1984). A further note on archaeological fly puparia. *Circaea* **2**, 103–105.

Pryor, L.D. & Boden, R.W. (1962). Blowflies as pollinators in producing *Eucalyptus* seed. *Australian Journal of Science* **24**, 326.

Putman, R.J. (1983). *Carrion and Dung: the Decomposition of Animal Wastes.* Institute of Biology, Studies in Biology No. **156**. London: Edward Arnold (Publishers) Ltd.

Ratcliffe, F.N. (1935). Observations on the sheep blowfly (*Lucilia sericata* Meig.) in Scotland. *Annals of Applied Biology* **22**, 742–753.

Reiter, C. (1984). Zum Wachstumsverhalten der Maden der blauen Schmeissfliege *Calliphora vicina. Zeitschrift für Rechtsmedizin,* **91**, 295–308.

Richardson, P.R.K. (1980). *The Natural Removal of Ungulate Carcases, and the Adaptive Features of the Scavengers Involved.* M.Sc. thesis, University of Pretoria.

Ring, R.A. (1967). Maternal induction of diapause in the larva of *Lucilia caesar* L. (Diptera: Calliphoridae). *Journal of Experimental Biology* **46**, 123–136.

Roberts, M.J. (1972). The structure of the mouthparts of some calypterate dipteran larvae in relation to their feeding habits. *Acta Zoologica* **52**, 171–188.

Roberts, R.A. (1933). Activity of blowflies and associated insects at various heights above the ground. *Ecology* **14**, 306–314.

Rognes, K. (1991). *Blowflies (Diptera, Calliphoridae) of Fennoscandia and Denmark.* Fauna Entomologica Scandinavica **24**. Leiden: E.J. Brill/Scandinavian Science Press Ltd.

Salt, G. (1932). The natural control of the sheep blowfly, *Lucilia sericata*, Meigen. *Bulletin of Entomological Research* **23**, 235–245.

Saunders, D.S. (1987). Maternal influence on the incidence and duration of larval diapause in *Calliphora vicina*. *Physiological Entomology* **12**, 331–338.

Schoof, H,F. & Mail, G.A. (1953). Dispersal habits of *Phormia regina* in Charleston, West Virginia. *Journal of Economic Entomology* **46**, 258–262.

Schoof, H.F. & Savage, E.P. (1955). Comparative studies of urban fly populations in Arizona, Kansas, Michigan, New York and West Virginia. *Annals of the Entomological Society of America* **48**, 1–12.

Sherman, R.A. & Pechter, E.A. (1988). Maggot therapy: a review of the therapeutic applications of fly larvae in human medicine, especially for treating osteomyelitis. *Medical and Veterinary Entomology* **2**, 225–230.

Shura-Bura, B.L., Shaikov, A.D., Ivanova, E.V., Glazunova, A.Ia., Mitruikova, M.S. and Fedorova, K.G. (1958). The character of dispersion from the point of release in certain species of flies of medical importance. *Entomologiskoe Obozrenie* **37**, 336–346.

Smith, K.G.V. (1956). On the Diptera associated with the stinkhorn (*Phallus impudicus* Pers.) with notes on other insects and invertebrates found on this fungus. *Proceedings of the Royal Entomological Society of London (A)* **31**, 49–55.

Smith, K.G.V. (1986). *A Manual of Forensic Entomology.* London: British Museum (Natural History) and Cornell University Press.

Smith, P.H. (1983). Circadian control of spontaneous flight activity in the blowfly, *Lucilia cuprina*. *Physiological Entomology* **8**, 73–82.

Southwood, T.R.E. (1978). *Ecological Methods.* London: Chapman & Hall.

Spradbery, J.P. (1979). The reproductive status of *Chrysomya* species (Diptera: Calliphoridae) attracted to liver-baited blowfly traps in Papua New Guinea. *Journal of the Australian Entomological Society* **18**, 57–61.

Stewart, M.A. & Roessler, E.B. (1942). The seasonal distribution of myiasis-producing Diptera. *Journal of Economic Entomology* **111**, 527–528.

Sychevskaya, V.I. (1962). Changes in the diurnal composition of species of synanthropic flies in the course of the season. *Entomologicheskoe Obozrenie* **41**, 545–553.

Thomson, A.J. & Davies, D.M. (1973*a*). The biology of *Pollenia rudis,* the cluster fly (Diptera: Calliphoridae). I. Host location by first-instar larvae. *Canadian Entomologist* **105**, 335–341.

Thomson, A.J. & Davies, D.M. (1973*b*). The biology of *Pollenia rudis,* the cluster fly (Diptera: Calliphoridae). II. Larval feeding behaviour and host specificity. *Canadian Entomologist* **105**, 985–990.

Thomson, A.J. & Davies, D.M. (1974). The biology of *Pollenia rudis,* the cluster fly (Diptera: Calliphoridae). III. The effect of soil conditions on the host-parasite relationship. *Canadian Entomologist* **106**, 107–110.

Vogt, W.G. & Woodburn, T.L. (1980). The influence of temperature and moisture on the survival and duration of the egg stage of the Australian sheep blowfly, *Lucilia cuprina* (Wiedemann) (Diptera: Calliphoridae). *Bulletin of Entomological Research* **70**, 665–671.

Watts, J.E. & Merritt, G.C. (1981). Leakage of plasma proteins onto the skin surface of sheep during the development of fleece-rot and body strike. *Australian Veterinary Journal* **57**, 98–99.

Wells, J.D. & Greenberg, B. (1992). Interaction between *Chrysomya rufifacies* and *Cochliomyia macellaria* (Diptera: Calliphoridae): the possible consequences of an invasion. *Bulletin of Entomological Research* **82**, 133–137.

Whiting, A.R. (1967). Biology of the parasitic wasp *Mormoniella vitripennis* (*Nasonia brevicornis* (Walker)). *Quarterly Review of Biology* **42**, 333–470.

Williams, C.B., Common, I.F.B., French, R.A., Muspratt, V. & Williams, M.C. (1956). Observations on the migration of insects in the Pyrenees in the autumn of 1953. *Transactions of the Royal Entomological Society of London* **108**, 385–407.

Williams, H. (1984). A model for the aging of fly larvae in forensic entomology. *Forensic Science International* **25**, 191–199.

Williams, H. & Richardson, A.M.M. (1984). Growth energetics in relation to temperature for larvae of four species of necrophagous flies (Diptera: Calliphoridae). *Australian Journal of Ecology* **9**, 141–152.

Williams, R.W. (1954). A study of the filth flies in New York City, 1953. *Journal of Economic Entomology* **47**, 556–563.

Willmer, P.G. & Unwin, D.M. (1981). Field analyses of insect heat budgets: reflectance, size and heating rates. *Oecologia (Berlin)* **50**, 250–255.

Wobeser, G. & Galmut, E.A. (1984). Rate of digestion of blowfly maggots by ducks. *Journal of Wildlife Diseases* **20**, 154–155.

Zeuner, F.E. (1941). The Eocene insects of the Ardtun Beds, Isle of Mull, Scotland. *Annals and Magazine of Natural History* **11**, 82–100.

Zinovjeva, K.B. (1980). Inheritance of larval diapause in crosses between two geographical forms of *Calliphora vicina* R.-D. (Diptera, Calliphoridae). *Entomologicheskoe Obozrenie* **59**, 498–509.

Zumpt, F. (1965). *Myiasis in Man and Animals in the Old World.* London: Butterworths.

Index